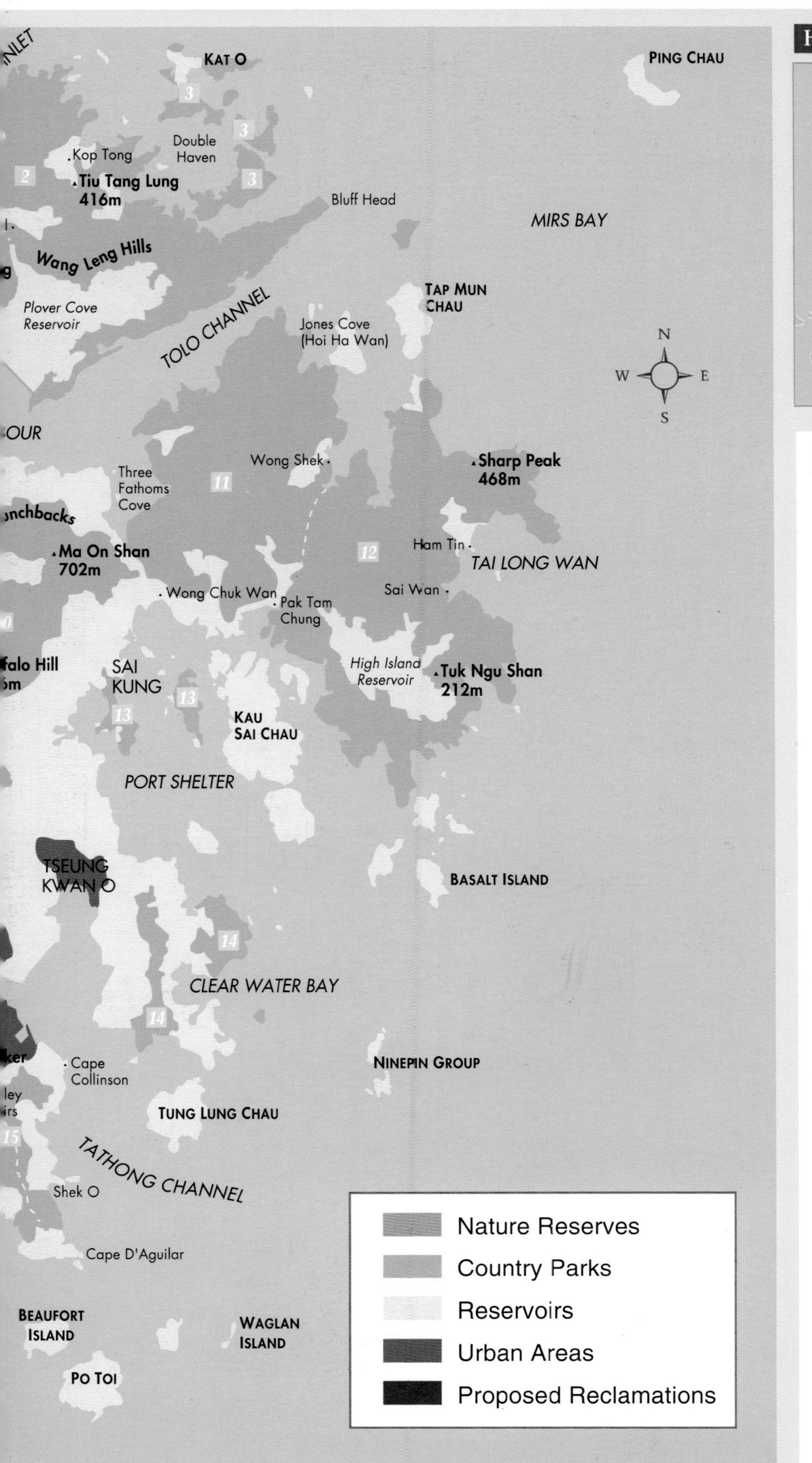

HONG KONG HARBOUR (*before 1887*)

KOWLOON

Stonecutters Island

Victoria Harbour

HONG KONG ISLAND

COUNTRY PARKS & NATURE RESERVES

1. Pat Sin Leng Country Park
2. Plover Cove Country Park
3. Plover Cove (Extension) Country Park
4. Lam Tsuen Country Park
5. Tai Lam Country Park
6. Tai Mo Shan Country Park
7. Shing Mun Country Park
8. Kam Shan Country Park
9. Lion Rock Country Park
10. Ma On Shan Country Park
11. Sai Kung West Country Park
12. Sai Kung East Country Park
13. Kiu Tsui Country Park
14. Clearwater Bay Country Park
15. Shek O Country Park
16. Tai Tam (Extension) Country Park
17. Tai Tam Country Park
18. Aberdeen Country Park
19. Pok Fu Lam Country Park
20. Lantau North Country Park
21. Lantau South Country Park

---

1. Futien Nature Reserve
2. Mai Po Marshes Nature Reserve
3. Kadoorie Farm and Botanic Gardens
4. Tai Po Kau Nature Reserve

HONG KONG'S WILD PLACES

An Environmental Exploration

To Derek.

Kelvin, Clare
& Nicola
X'mas 95

HONG KONG'S WILD PLACES

AN ENVIRONMENTAL EXPLORATION

Edward Stokes

HONG KONG
OXFORD UNIVERSITY PRESS
OXFORD NEW YORK
1995

Oxford University Press

Oxford New York
Athens Auckland Bangkok Bombay
Calcutta Cape Town Dar es Salaam Delhi
Florence Hong Kong Istanbul Karachi
Kuala Lumpur Madras Madrid Melbourne
Mexico City Nairobi Paris Singapore
Taipei Tokyo Toronto

and associated companies in
Berlin Ibadan

Oxford is a trade mark of Oxford University Press

First published 1995
This impression (lowest digit)
1 3 5 7 9 10 8 6 4 2

Published in the United States
by Oxford University Press, New York

British Library Cataloguing in Publication Data
available

Library of Congress Cataloging-in-Publication Data
Stokes, Edward, 1948–
Hong Kong's wild places: an environmental exploration/Edward Stokes.
p. cm.
Includes bibliographical references (p.) and index.
ISBN 0-19-586601-0
1. Man—influence on nature—Hong Kong. 2. Landscape assessment—Hong Kong
3. Environmental degradation—Hong Kong. 4. Conservation of natural resources—Hong Kong.
5. Hong Kong—Environmental conditions. I. Title.
GF658.S76 1995
333. 78'2'095125—dc20 95-24786
CIP

Printed in Hong Kong

Published by Oxford University Press (China) Ltd
18/F Warwick House, Taikoo Place, 979 King's Road,
Quarry Bay, Hong Kong

Wherein lies the reason that
a good man so much loves landscapes?
'It is because the message of the forests and the streams,
the companionship of the mists and clouds,
are always in his dreams.'

KUO HSI, SONG DYNASTY

*Increasingly modern man cannot face ancient
and untamed nature, where nature itself, the elements,
rule, not man; where iron and concrete and plastic are
powerless; where [nature] defeats all use except
enjoyment on its own terms. That is why we have
lost so much and are now lost ourselves.*

JOHN FOWLES, 1989

To all those striving to preserve
Hong Kong's natural heritage

THE SWIRE GROUP
SPONSORED THE WORK FOR THIS BOOK

PREFACE

My attraction to wild country was born in Hong Kong, where I grew up, and confirmed when I later returned to Australia, my homeland. There, awed by the outback's space and light, I grew to love its harsh grandeur.

The spirit of the past seemed often to colour the country; the land, I came to see, told its own story. Long-gone people had affected the natural landscape. Some of their changes to the environment could barely be seen, others were clearly evident, and some were glaring and troubling. All, to varying degrees, could be documented and photographed.

Experiences in Australia thus shaped my thoughts, but the Hong Kong landscape lay deeper in my memory. Vivid images of my boyhood here periodically surfaced: teenage rambles past gullies cascading after typhoon rains; winter hikes through dry gravelly hills; and boat expeditions towards islands that then seemed mythically distant. Always there were peaks and islands. And always, taken for granted, were the relatively unpolluted sea and sky of Hong Kong in the 1950s and 1960s.

Its mountains and valleys in mind, a few years ago I visited Hong Kong. I hiked across Hong Kong Island, from Wong Nai Chung Gap, past Repulse Bay, and round to Tai Tam. Gazing from Repulse Bay's encircling hills—past dominating tower blocks, down to where the original Repulse Bay Hotel once nestled, down to the 'Butterfly Valley' of my childhood—was to mourn.

Countless other Hong Kong hillsides, of course, had been turned to concrete, many to house the post-war millions. In the New Territories—the agricultural lowlands of the 1950s and 1960s—summer-wet paddy fields and winter stubble had almost all gone. Satellite cities stood where only decades before buffalo had worked and wallowed.

But in Hong Kong's extensive uplands and remote corners much beautiful country still remained. Some of it was truly wild and majestic, its craggy peaks and plunging valleys as dramatic and beautiful as ever. Indeed, the paradoxical proximity of Hong Kong's empty hills and crowded streets was even more striking than before.

It was time, I decided after my visit, to re-explore Hong Kong's remaining countryside, to climb again into its wild places. I returned here in 1993. The plane descended over sinuous Lamma, down past Tai Mo Shan's massive buttresses, lower still opposite the jagged summit of Lion Rock. Then, even before the wheels grounded, came the unmistakable stench of the Kai Tak 'nullah'. It was mid-May: hot and humid.

The popular interest in Hong Kong's countryside is relatively recent. Despite that, many books have catalogued specific aspects of its natural heritage, and a few have given more rounded impressions. None of these, however, presented a general account of Hong Kong's country as I now saw it: a magnificent, constantly changing, now severely threatened stage where, through millenia, nature and man have affected one another. Thus this environmental history—an exploration of Hong Kong's endangered countryside.

The book aims to record the changes in the natural landscape, and the consequences—both good and bad—of human actions there. In doing so, it presents an overview of Hong Kong's geography, and of the ecological value of its remaining countryside. Less happily, the book also points to the degradation and pollution that have overtaken urban Hong Kong in recent decades—and which now also spread into the country areas.

The photographs were almost all taken in Hong Kong's Country Parks. Above all, they celebrate the Territory's beauty and grandeur. But they also aim to reveal past human impacts that can still be seen in the natural environment—and, like the text, they warn of the threats that now face the countryside.

The book is structured around alternating 'historical' and 'regional' chapters. The former set the scene, narrating the environmental changes—both natural and human-derived—during successive periods. The latter are journeys across the country, giving impressions of Hong Kong's distinct regions.

In the early chapters, 'Hong Kong' refers to the mainland and islands that now comprise the Territory. The spelling of local place names varies widely. I chose to follow those used on the government *Countryside Series* maps; I have also mostly followed the maps' preference for either Chinese or English names, where both exist. Names of places in China are in pinyin. I usually hiked and photographed alone, but sometimes with friends, hence both 'I' and 'we' appear in the text.

Most writing about Hong Kong's natural heritage is by Europeans or Western-educated Chinese, but where possible I have emphasized local Chinese aspects. For those interested, the 'Further Reading' section includes a bibliography and notes on how I researched and referenced the text. I also hope that the Hiking and Conservation Notes, and the Photographic Notes will encourage readers to explore Hong Kong's countryside themselves.

Although widely shared by many people concerned about Hong Kong's natural heritage, the views expressed in this book are my own.

Edward Stokes
Lamma Island, Hong Kong
May 1995

ACKNOWLEDGEMENTS

My ideas were much influenced by Fernand Braudel's and Lyman van Slyke's historical geographies. These wide-ranging books gave fresh insights, especially concerning historical and geographical 'space' and 'time'. I also owe a debt to some Hong Kong writers, whose accounts of the local countryside have been most valuable to a non-scientist. G. A. C. Herklots and G. S. P. Heywood both documented the countryside of the 1930s; Stella Thrower made a study of the Country Parks in the 1980s; and David Dudgeon and Richard Corlett recently wrote an ecological study of Hong Kong. Austin Coates and James Hayes, while only partly concerned with the environment, inspired me through their impressions of post-war Hong Kong's rural life.

My late father knew of my plans for this book, and his love of the Hong Kong countryside remains with me. Gwenneth Stokes, my mother, contributed much to the work through her knowledge of Hong Kong—and of grammar! In Hong Kong, friends, mostly Lamma-ites, urged me on: Rob Baxter, Kevin Bishop, Don Brech, Ginny Davies, Ian Driscoll, Roz Forestal, Nigel and Candy Hicks, Bill Leverett, Madeleine Lynn, Helen Ma, Jenz-Peter Mücke, David St Maur Shiel, Sally Trainor, and Evans Ward. Kevin Bishop was always ready to come searching if I failed to return from lone hikes. Barbara Mobbs, my agent in Australia and herself once a Hong Kong 'belonger', also helped.

My research was done mostly at the University of Hong Kong, where I owe numerous debts. Special thanks go to Dr Elizabeth Sinn, who encouraged throughout and who commented on the text; to Dr John Hodgkiss, who gave me insights into the Country Parks; and to Dr Richard Corlett and Dr David Workman who read parts of the text. At the Special Collections of the University of Hong Kong Libraries, the Librarian Mr Y. C. Wan offered much assistance, as did Mr C. H. Mak, Librarian of the City Hall Reference Library. The Centre of Asian Studies provided interesting and thought-provoking seminars.

Hong Kong's main environmental groups have helped, especially World Wide Fund for Nature (WWF) and Friends of the Earth. My special thanks go to David Melville, Executive Director of WWF and Dr Lew Young, Manager of WWF's Mai Po Marshes Nature Reserve, who both gave valuable advice; Ken Chu of WWF also helped. Simon Chau, then with Green Power, commented on the text. Dr Gary Ades, Senior Conservation Officer at Kadoorie Farm and Botanical Gardens, advised on the text's ecological aspects. Professor Brian Morton, at the Swire Institute of Marine Science, Mr C. W. Lai, at the Country Parks Division of the Agriculture and Fisheries Department, and staff at the Royal Observatory helped with my field work and photography. Mr Lai also helped with parts of the text. Legislative Councillor Christine Loh gave much encouragement to conservation projects stemming from this book.

Developing and producing this book has been a team effort, and my special thanks go to the Academic and General Unit of Oxford University Press (China) Ltd for their much appreciated encouragement and ideas.

My final, and greatest, acknowledgement goes to the Swire Group, who sponsored my work, and without whose generous support this book might not have been completed. My lasting appreciation goes to Sir Adrian Swire and Sir John Swire, who first recognized the project's potential. I also thank Glen Swire and Charlotte Havilland who helped in London. In Hong Kong, Nick Rhodes has greatly encouraged the work, and, together with his staff at Swire Public Affairs, offered much assistance and support.

CONTENTS

Seen across the West Lamma Channel, Hei Ling Chau and Sunshine Island frame Lantau Island.

INTRODUCTION

From the tops of the mountains the view is grand and imposing in the extreme. Mountain is seen rising above mountain, rugged, barren and wild.

So wrote the botanist Robert Fortune of Hong Kong's landscape as he saw it in 1844. A century and a half later his words still hold true for many parts of the Territory. While exploring Hong Kong for this book I was often awed by panoramas of peaks and valleys, bays and islands. I saw how beautiful and varied Hong Kong's countryside is—but also how encroaching development severely threatens it.

This book, by setting Hong Kong's contemporary countryside in its historical context, seeks to emphasize three themes: the great beauty of the Territory's natural landscape; the severe threats now facing its countryside; and the irredeemable folly of destroying our natural heritage.

The common Chinese expression for 'landscape' is *shan shui*, or 'mountains water'. In few other places are mountains and water blended in such dramatic grandeur as in Hong Kong. Fewer still have wild country so close to a dynamic, glittering city.

Hong Kong's precipitous terrain has kept its people on the lowlands, a mere fraction of the Territory's area. Today 95 per cent of the population live on less than 20 per cent of Hong Kong's 1,076 square kilometres. By contrast, most of the upland country lies within twenty-one Country Parks. These natural oases cover more than 40 per cent of the total land area—a very high proportion by any standards, and still more so given Hong Kong's size and population. The Country Parks offer a stupendous contrast to urban Hong Kong. They hold great ecological diversity, historical interest, recreational value, and even solitude.

Historians, Fernand Braudel wrote metaphorically, tend to 'linger over the plain' often unwilling 'to approach the high mountains nearby'. He continued:

Yet how can one ignore these conspicuous actors, the half-wild mountains.... How can one ignore them when often their sheer slopes come right down to the sea's edge?

'A historian needs a good pair of boots', an Australian writer once noted. Hong Kong wore out two pairs of mine. Hiking and photographing in the rugged uplands and along the coasts brought both charms and challenges. But it was there, during days and nights spent in the wild, that Hong Kong's natural landscape came to life for me. Treading along centuries-old boulder pathways, searching around overgrown ruins, imagining the toil of building hillside terraces, I saw, and sometimes felt, the past. As Lyman van Slyke put it, 'remembered time lay like geological strata on the land'.

The view that Hong Kong is merely an urban phenomenon—a place born of British rule and Chinese enterprise, without roots in the distant past and in its countryside—seemed increasingly flawed. A central theme of this environmental exploration is that Hong Kong's natural and human origins lie much further back. Indeed, G. B. Endacott's statement, concerning Hong Kong Island, that its history 'really begins with the coming of the British' reflects only one 'reality'—that of colonial, urban Hong Kong.

Like Braudel's and Lyman van Slyke's historical geographies, this environmental exploration describes the interplay between 'natural' and 'social' events. Throughout, the natural order infinitely slowly evolves, while year-to-year nature repeats itself: the monsoons alternate, plants grow, animals mate and die. People arrive and begin altering—sometimes improving, sometimes degrading—the natural landscape. Environmental change is a constant: only its degree and speed vary.

How have nature's cycles and human intervention affected Hong Kong's 'natural landscape'—as I take it, the entire ecological complex, the stage for man's association with 'the earth'? And how has the local landscape shaped Hong Kong's people? The story falls into six periods.

In the first, Hong Kong's environmental change was so gradual that only a drastic telescoping of aeons indicates any alteration at all. Beginning with the creation and shaping of Hong Kong's rocks and landforms, continuing through the later colonization by plants and animals, the period ends about 5000 BC when the sea level stabilized. The landscape was then virtually what we see today, though covered mostly with forests.

The second period covers the seven thousand years of human settlement from some time after 5000 BC until 1841. The prehistoric population was insignificant, but later, much more numerous Chinese settlers greatly altered the country. From our frenetic standpoint this rural era, patterned by tradition and superstition, saw only gradual environmental change. However, over about the last ten centuries the natural forests were mostly

destroyed and virgin land was ploughed into fields. Many of the larger native animals lost their habitats and were hunted out.

Some sixty years cover the third period, the years from British colonization in 1841 until 1900. After 1841 an urban trading centre developed very rapidly on Hong Kong Island. Its fast growing population placed ever-increasing demands on the Island's countryside. The granite hills were quarried for buildings, the harbour's natural shoreline was built over. Later, villages and fields were submerged under valley reservoirs and the 'barren rock' was steadily reforested by man.

The fourth period extends from 1900 until the Second World War engulfed Hong Kong in 1941. This was a period of general progress: typhoon shelters were built, roads and a railway cut across the rugged hinterland, more reservoirs were completed and reforestation spread. Easier access into the countryside encouraged a strong appreciation of Hong Kong's natural heritage among hikers and naturalists. But in the New Territories, which had been ceded in 1898, the agricultural lowlands and uplands remained largely unchanged.

The forty-five years since the Second World War saw Hong Kong transformed by human energy and modern technology. With dizzying speed hillsides became terraced concrete, bays disappeared under vast reservoirs, massive reclamations recast the coastline. Never before had the natural setting been so drastically changed. Indeed, the population growth was so extreme and the development so dramatic—and later damaging—that the mere forty-five years since 1945 form two distinct environmental periods.

From 1945 until about 1970, post-war reconstruction and the influx of refugees, bringing new housing and industry, laid the foundations of modern Hong Kong. As development gathered pace the urban fringes began reaching into and consuming the countryside. Yet remarkably, especially early in this period, the New Territories remained mostly agricultural. Meanwhile hillsides that were denuded during the war were replanted with trees, and more reservoirs were built.

The sixth period is from about 1970 to 1990. Driven by affluence and technological power, combined developments swept away in less than a generation Hong Kong's agricultural lands and traditions that reached back at least one thousand years. Meanwhile pollution escalated alarmingly, and 'New Towns' spread across the rural lowlands. One ray of hope stood out: the Country Parks, established in the late 1970s to help preserve the remaining countryside. Sanctuaries of untrammelled wilds, the Country Parks were protected by legislation—and, in part, by their rugged, challenging terrain.

Ever since 1841, and especially during the later twentieth century, a steadily closer relationship evolved between country and city. Hong Kong's inexorable urban growth dictated its countryside's fate: like it or not, the country learnt to listen for the city's rumblings. Troublesome or worse for the countryside and its villagers, this interplay does make for a more multi-faceted environmental saga than otherwise—for to write about Hong Kong's countryside without also outlining its urban expansion would be to ignore the underlying forces.

Hong Kong's ever-growing population and trade have always been central to this inter-play. The city-based forces in the early colonial period merely modified conditions in the nearby countryside. Today's urban-economic forces, however, threaten the very survival of the countryside—as the development of the new airport at Chek Lap Kok indicates.

This book's focus is on how people have changed the natural environment, but it would be hardly human to ignore how the land itself has affected Hong Kong people. So the story also touches on how the terrain, climate, and weather have moulded people's ordinary lives and spiritual beliefs.

Hong Kong's mountain lookouts dramatically reveal the beauty of the surrounding uplands. But they also emphasize the threats facing the Country Parks today. Atop numerous summits I have revelled in the wild grandeur around me—and looked down on degradation and pollution spreading across the lowlands, around the coasts, through the air.

Six environmental periods are sketched above. But there is a seventh: the present and future. Shenzhen, as seen from peaks in the northern New Territories, is awesome: a wall of tower blocks now lines the Shenzhen River. With Hong Kong's future inextricably linked to the unbridled economic development across the border, the Territory's natural environment now faces perhaps its greatest challenge. Where will the concrete end?

The human history of Hong Kong, even more its period as a British colony, fades into insignificance beside the infinitely longer story of 'natural' Hong Kong. Yet here especially people often seem blinded by a single reality—'now'—a mere moment in a human-environmental saga that reaches into the distant past and into the unknown future.

Should we most admire Hong Kong's buildings, or the hills that provide their setting? Should we cherish its spreading reclamations or its natural waterways? Should we be concerned about the natural environment that our generation will bequeath? This book raises many environmental issues, but offers no easy answers. Whether Hong Kong's remaining countryside is preserved or destroyed depends on the community's values—and on government, business, and individual choices.

If this book challenges Hong Kong's old environmental attitudes it will have succeeded. If it also inspires people—and you, the reader—to become actively involved in promoting conservation it will have succeeded doubly. The more we venture into the Territory's countryside—hiking through it, sensing its rhythms, observing its changes—the greater the chance of Hong Kong's wild places actually being preserved. We all share responsibility.

The summer monsoon has a major impact on Hong Kong's landscape. Here, at dawn, towering summer clouds rise over Lamma Island.

HISTORY

PRIMEVAL LANDSCAPE

The sand comes and goes, wave after wave.
The surge dies, another rises.
Stirring and re-forming endlessly,
They level the mountains and seas in time.

PO CHU-I, TANG POET

Rain sometimes fell as misty drizzle, sometimes as showers from passing cloudbursts. And sometimes the land was deluged by torrential downpours: incessant rain that left the hills and valleys sodden; that turned streams to raging torrents; that penetrated every nest and hide. Animals sniffed the ground, their instinct merely recording—*wet*!

No one remarked that the summer monsoon had set in, or compared the rainfall with previous years. For the time is about 5000 BC, and, so far as we know, humans had not yet reached Hong Kong.

Every year the spring mists came and went, replaced by clear skies and oppressive heat. At last towering rain clouds rolled in off the sea, promising to break the sultry weather. Low and threatening, dark clouds veiled, then hid the hilltops. Thunder rumbled. Rain began falling—lightly, then more heavily, finally with a leaden hammering. The world dissolved into greys: lightning flashes cut through the gloom, and thunder cracked overhead.

The elements almost spent, there came a lull. The thunder faded, rain fell more gently. One hilltop, then another, appeared—dull green, saturated, wreathed in vapours. The hills ran with water: countless streams swept downwards into gullies where torrents leapt and surged.

So, over countless centuries, the summer monsoon rains returned. Life-giving but sometimes lethal, they wore down the land but refreshed plants and animals.

About seven thousand years ago—5000 BC—the terrain was much the same as today's, though the land was thickly forested. But about seventeen thousand years ago—15,000 BC—the scene was very different. Then, with the sea level much lower than today, Hong Kong's future territory was simply a tangle of inland hills rising above a wide coastal plain. Yet even that landscape was dramatically transformed from what it had been before—long, long before.

So how and when did the peaks, valleys, and coasts of today's Hong Kong take shape? When did plants and animals first colonize the area? And how did the surrounding seas affect the land?

Hong Kong, geologically and climatically, is part of South China. Its landforms are an extremity of Guangdong's—and they form the province's most complex, striking coastline. East of where today the Pearl River enters the sea, sediments were laid down beneath shallow waters during Devonian times, about 360 million years ago. These silts, slowly compressed to sandstones, shales, and conglomerates, are Hong Kong's first rocks—seen today at Bluff Head near Tolo Channel.*

During the middle Jurassic period, between about 170 and 150 million years ago, the region had numerous seething vents. Volcanoes spewed out lava, and incandescent eruptions blasted tonnes of ash, stones, and boulders into the air. Amid sulphurous fumes, beneath dust-hazed skies, the volcanic debris fell, settled, cooled, and solidified into igneous rocks that covered the landscape. Massive pressures then folded these rocks during the Upper Jurassic period, forcing them up into the roughly north-east to south-west trending ranges that frame the landscape today. No sooner were these volcanic uplands formed than their weathering and subsequent erosion began.

As the Jurassic period drew to a close, about 165 to 140 million years ago, more molten rock welled up in the depths. This magma,

* *This treatment of the geological formation of the Hong Kong region is extremely cursory. The geological processes described in the text were separate from the infinitely slow 'drift' of the continents due to 'plate tectonics'.*

forced upwards by immense pressure, thrust itself into the region's older sedimentary and volcanic rocks but did not reach the surface. It gradually cooled and solidified underground, forming vast masses of granite. Then, from about 135 to 100 million years ago, the land was relentlessly worn away—and ridges, spurs, and valleys were gradually sculpted from their parent rock. Over wide areas of Hong Kong the sub-surface granite was exposed by denudation of its cover. Then, during the Upper Cretaceous period, about 100 to 90 million years ago, the area was once again massively transformed, folded, and twisted by earth movements.

The last 60 million years have seen a continuing gradual wearing down of earlier formations. Slowly the modern landscape appeared: mountains took on distinct features, valleys were widened, and some limited alluvial plains formed. By Quaternary times, two million years ago, Hong Kong's topography was largely as today, except the sea was much lower.

Among the most important early organisms were those that lived in earth. Surface weathering created a layer of waste mantle (or disintegrated rock). But it was the endless cycle of plants growing and dying, and the waste from detritivores (animals that eat dead plants) that transformed the waste mantle into nutrient-rich soils. These tiny creatures, by initially breaking up, aerating and enriching the waste mantle, created soils rich enough for plant species to evolve and spread.

Knowledge of Hong Kong's early organisms is sketchy, but some of the detritivores seen today are their descendants. Among them are the beetles, bugs, cockroaches, termites, snails, worms, and woodlice. The Wingless Litter Cockroach is one of these detritivores: dark brown and four centimetres long, its sole diet is dead leaves.

Sea coasts are always regions of dynamic change, moulded by the winds, waves, and tides. However, the outline of Hong Kong's coastline reflects terrestrial (or land) forces, not marine (or sea) ones—for, until quite recently, the coast we know did not even exist. Indeed, the present coastline, steeply sloping and deeply indented, is a classic ria (or submerged) coast.

During the last two million years sea levels have varied by as much as 150 metres. Global earth movements contributed to these fluctuations, but the major cause was the cooling and warming of

By about two million years ago, Hong Kong's topography was largely as it is today: rugged peaks, such as Lion Rock, dominated the scattered lowlands.

Plants most easily colonize sheltered, moist areas where they take root even in minute rock fissures.

the global climate. During the Ice Ages vast amounts of sea water, held as continental ice, lowered the world's sea levels—which rose as the ice melted during warmer periods.

Late in the last Ice Age, about 15,000 BC, the sea level was some 100 metres lower than today. Around Hong Kong, streams that rose in the hills joined to form long rivers, which meandered for some 200 kilometres across coastal plain to the sea.

Then, from about 15,000 BC until about 5000 BC, the global climate warmed, the ice sheets slowly melted, and the sea rose. The coastal plain flooded first. By about 10,000 BC, when the worst of the Ice Age cold had gone, the coastline was probably near the Dangan Islands (about 30 km south of Hong Kong Island). Later, as if the Pearl River were sucking the sea into its mouth, the waters advanced towards Hong Kong: valleys became bays, ridges became headlands, peaks became islands. The sea finally stabilized at its current level about 5000 BC.*

Partly because of the gentle slope of the coastal plains, this sea invasion had been extremely rapid, lasting a mere 10,000 years or so—in geological time spans, a mere moment. Evidence of the sea flooding around Hong Kong is seen in the land fossils found under the sea, and in the silt channels lying across the seabed—the remains of the Ice Age coastal rivers.

So today we see a ria coastline: a coast whose general shape was formed by much earlier terrestrial processes and later delineated by the sea. But what of the surging ocean, and the silt-laden Pearl River? Marine erosion and deposition have little to say concerning the general outline of Hong Kong's coast, but they say much about the later refinement of the coast's details.

Marine erosion is most evident along Hong Kong's eastern coasts, and especially among the eastern islands. Crashing swells there have cut back the rock into cliffs, platforms, and caves. As on land, erosion removed the softer rock first, while the hardest withstood the ocean onslaught longest. Hence, on a small scale, the less uniform the original rocks the more jagged the scene today. Rain water and salt spray, by weathering and so weakening the coastal rock, hastened this ocean erosion.

Around Hong Kong's western coasts the story is one of silt, not spray. The Pearl River, which drains vast areas of South China, has always been the central actor. In recent times the river has deposited some 80 million tonnes of silt into its estuary each year (the Kai Tak airstrip, by comparison, took about 16 million tonnes of earth and rock). The Pearl River's ancient silt load would have been less than today's, as South China was then better forested; none the less, the river must still have brought down vast quantities each year.*

As a result Deep Bay, shallow enough even after the sea level rose, gradually silted up and became shallower still. The river silts settled around the shore of Deep Bay, and the coast extended itself outwards metre by metre (Deep Bay's name refers to its shape, not its depth).

The Pearl River was not the only source of silt. Much of Hong Kong is dominated by granite rocks—and these, combined with the post-Ice Age climate's heavy tropical rainfall, formed a virtual silt factory.

Unweathered granite has countless fractures, hairline cracks that allow deep penetration by water and by the natural acids formed when rain absorbs atmospheric gases, especially carbon dioxide. The acidic solutions slowly weather (or decompose) the granite; at the surface, the loose material is eroded to leave boulders or large 'tors', remnants of granite corestones (or solid blocks). Hence Hong Kong's boulder-strewn slopes.

The process continues today. Hong Kong's sometimes tragic landslides, often caused by gravity acting on sodden, deeply weathered slopes, are one result: yellow-brown slashes across hills, a mess of mud crashing suddenly downhill. Landslides thus partly

* *After about 5000 BC the sea level rose a little higher, before falling back to today's level as the global climate cooled marginally. Local 'raised beaches', old beaches that now lie above the current sea level, are evidence of this.*

* *The Pearl River (Zhujiang) has three main tributaries: East River (Dongjiang), North River (Beijiang) and West River (Xijiang). Together they drain some 425,000 square kilometres of Guangdong. The Pearl River delta is sheltered from the ocean by numerous offshore islands, so hastening the processes of alluvial deposition.*

By about 10,000 BC, rising sea levels had brought the coastline near to this island —one of the Dangan group, about 30 kilometres south of Hong Kong Island.

shaped the slopes we see, bringing down material that spread out across alluvial valleys, formed coastal deposits behind sandbars, and spilled into the sea.

With the last Ice Age finally ended, Hong Kong emerged in its 'modern' guise. In a real sense it was modern. For—had people not subsequently radically altered the terrain, flora, and fauna—the natural landscape of about 5000 BC is roughly what would still exist today.

How could 'modern' Hong Kong be pictured? What were the main features of its topography, climate, plants, and animals some seven thousand years ago?

The landscape, now fringed by sea, had great variety in its relief and shape. There were rugged uplands, sheltered lowlands, and a coast of extraordinary complexity. Almost everywhere distinctly different topography existed side by side.

The uplands had by far the most dominating aspect. The highest points were impressive mountains, whose ridges and foothills extended like tentacles across the landscape. In the north-west, a large alluvial plain spread across the mainland, rising just fifteen metres above the sea. But elsewhere the lowland valleys were crowded between plunging slopes. The coastal strip was, in most places, precisely that—and often there was no coastal strip at all.

The mountains, their gentler foothills now submerged under the sea, showed only their highest—and so steepest—aspects. This was to have long-term repercussions. It meant flat, or even mildly sloping, land was desperately scarce. Allied with the deeply weathered granites, it meant landslides and hill collapses would always threaten. And, despite a maze of rocky streams, it meant there were no lowland lakes—and barely one respectable river.

But, though clearly challenging, the landscape was full of charm. Across a small archipelago numerous islands rose steeply from the sea. Along almost every section of the coast beaches, bays, coves, headlands, and peninsulas vied for attention—

Hong Kong's intense summer rainstorms are a major erosional force.
These Round Island boulders, however, have been smoothed by wave action not by rainfall.

Over aeons, water-borne rock particles, coursing downhill, have gouged out waterfalls, such as this one at Ng Tung Chai, north of Tai Mo Shan.

compelling details of a coastline both sinuous and long. In the hills watercourses cut ever deeper into their parent slopes. They sculpted gullies, ravines, and waterfalls which, after summer rains, fell in sparkling cascades.

Evidence of Hong Kong's climate before about 5000 BC is relatively sketchy. However, during Pleistocene times and until about 10,000 years ago, it was probably cooler and drier than today. Then, as the Ice Age receded, the weather slowly warmed. Between about 5000 and 4000 BC, the climate stabilized very close to today's levels, roughly when the sea level steadied. The two, of course, were connected.*

The global climate has changed only marginally since then. Indeed, the annual heating and cooling of the Asian land mass—the single most profound force governing Hong Kong's monsoons—extends back much further. So, despite some minor fluctuations, the local climate has probably changed little in the last 6,000 years. Temperatures were highest from about 5000 to 3000 BC, when most world regions were around one to two degrees Celsius warmer than today.

The future trading city's climate was clearly imported—twice each year. The air currents that drifted (and sometimes blasted) overhead came either from the steppes far to the north, or from the equatorial seas to the south. The climate was thus a mix of two opposing forces, the overlapping of temperate 'continental' and tropical 'oceanic' influences. One or the other source generally dominated, one or the other monsoon blew.

We can be certain that by about 5000 BC the winter, north-east monsoon brought dry, chilly, and sometimes cold air. In summer the south-west monsoon brought tropical heat, humidity, and rainfall. The vast bulk of the rain fell during the hot summer months, most of it in torrential deluges. Spring and autumn were

* *Some of the existing fauna suggest that Hong Kong's climate was previously more tropical than that encountered today. These purely tropical species include Burmese Pythons, Dog-faced Fruit Bats, Atlas Moths, Civets, and Pangolins.*

*Subtropical forest covered Hong Kong before the primeval forest was cut down.
In places, such as here in Kam Shan Country Park, mature woodlands have grown back today.*

unpredictable, and during spring jostling masses of cold and warm air created mists and fog.

A thickly forested landscape, could one visit the Hong Kong of about 5000 BC, would be the most startling difference. The grass-and-shrub hills that are so widespread today were yet to come. Indeed, far from being sparse, Hong Kong's primeval 'climax' vegetation was dense subtropical forest. It covered the valleys, climbed the slopes, and perhaps even reached the peaks.*

Long before, plant species had spread into the Hong Kong area from both the temperate north and the tropical south. As the post-Ice Age climate was gradually modified by warmer, wetter conditions, the region's older pine forests were steadily replaced by subtropical woodlands. This 'broad-leaved' evergreen forest spread, and by about 8000 BC was probably well-established throughout Hong Kong. That locally woodlands never developed into tropical rainforest, the usual 'climax' tropical vegetation, is explained by the cool, dry winters—and by typhoons which from time to time destroyed the largest trees.

The forest had a wide mix of species, dominated by the oak and laurel families. Perhaps ten or twenty species were seen commonly in particular areas, as is usual in tropical and subtropical forests. The woodland was especially lush and thick in the wetter, warmer lowlands; and rather less dense on the cooler, windswept uplands. Leaf litter broke down rapidly in the wet summer heat, forming rich surface composts above otherwise typically poor, rain-leached tropical soils.

The exact number of plant species that primeval Hong Kong supported is of course unknown. However, one can reasonably speculate that there were perhaps 1,500 species—as today's species numbers suggest. By 1910 botanists had described 1,580 local species; and Hong Kong now has some 2,000 plant species, including about 300 imported ones. That Hong Kong could support such a large floral diversity reflects its mixed climate.

In sheltered areas, lianas (tropical climbing plants) festooned an upper forest canopy that often shut out the sky. The understorey was fairly open, and along the ground were ferns, mosses, and lichens. Decaying leaves, twigs, fruits, fallen branches, and rotting trunks lay on the ground. The canopy filtered much of the light, but early and late in the day rays shafted into the forest. They shone brilliantly past tree trunks, etched the inner structure of leaves, and lit up a wonderland of greens.

** That forest may well have reached the peaks is suggested by the fact that, on the sheltered parts of some peaks today, trees and orchids typical of tropical forests can be seen.*

The region's animals, like its plants, reflected the mixed tropical-temperate climate. Indeed, as zoologists have since learned, Guangdong straddles a transitional zone between two distinct botanical and zoological regions: the tropical Oriental region and the temperate Palaearctic region.

Tropical forests typically have a wide diversity of fauna, but small populations of each species. So it was in the Hong Kong of about 5000 BC. Hence, when people subsequently cut down the forests, each species was particularly vulnerable and many disappeared.

The fauna possibly included elephants, and recent archaeological discoveries indicate that rhinoceroses lived here. It certainly included crocodiles, leopards, tigers, and other smaller mammals such as badgers, bats, civets, macaques, pangolins, porcupines, rats, shrews, and squirrels—and countless birds, reptiles, insects, and other invertebrates. The smallest creatures were no larger than a few pinheads, the largest were awesome. Some were carnivores, but herbivores were by far the most numerous.

Excluding elephants and rhinoceroses, about which little is known locally, the largest animals were Long-nosed Crocodiles, skeletal remains of which have been found near the Pearl River. The much larger Brackish-water Crocodiles may have lived here, but there is no firm evidence. South China Tigers and Leopards regularly wintered here, mostly preying on the larger mammals.

One of the tiniest Hong Kong animals, barely as thick as a big cat's whiskers and a few centimetres long, was the Iron Wire Snake. Of its forty-odd much larger cousins, the most venomous were two species of Kraits, two Cobras, two Vipers, and a Coral Snake—their deep colours signalling danger. The Chinese Cobra fed on reptiles, including other snakes, amphibians, and some smaller mammals. Burmese Pythons, up to six metres long, could kill virtually anything—even Barking Deer. Elusive and timid vegetarians, Barking Deer moved nimbly through the forests, browsing mostly on young shoots and fallen fruits.

Wild Boars rivalled Burmese Pythons for brute force. Large males weighed over 100 kilograms, though their curved tusks—actually extra-long canines—were used mostly for grubbing up roots. Burrow-dwellers often heard them shoving through the undergrowth: Chinese Porcupines (their genus name *Hystrix* surely conjures up quivering quills) rattled their tails when disturbed; and Chinese Pangolins rolled themselves into tight, scaley balls.

Crypsis, the use of shape or colour to avoid predators, was often invaluable—as butterflies, reptiles, and amphibians subtly showed. Romer's Tree Frog was brown-green and so minute—barely one centimetre across—that it has only recently been discovered. Short-legged Toads were as mottled-brown as the leaf litter they lived among, though not as cleverly concealed as the ludicrously elongated Stick Insects. Forest Geckos had bark-like colours and, better still, special toe pads for escaping up vertical rocks or trees. They shed their tails to confuse predators—and could later grow replacements!

Flight was another defence. Countless insects flitted and whirred: bees, beetles, crickets, cicadas, dragonflies, damselflies, earwigs, flies, grasshoppers, moths, mosquitoes, and termites. As it is today, Hong Kong was especially rich in butterflies, dragonflies, and damselflies, many of which were superbly coloured. Along with fruits, insects formed the main diet for many birds, the arrival of whose hatchlings coincided with the summer insect explosion. Black-eared Kites, Hong Kong's commonest bird of prey, soared and swooped—undisputed lords of the winter skies.

Sheer numbers could be a powerful defence, though this generally safeguarded the species rather than individuals. Atlas Moths, conspicuous Saturnids with superbly patterned ochre-and-orange wings twenty centimetres across, had a brief life cycle—but each female laid 100 or more eggs. Queen Common Wet-wood Termites lived for up to twenty years, amazingly producing some 10,000 eggs every day. Their flying progeny were preyed on in flight, or soon after shedding their wings. Yet, aided by social networks and feudal hierarchies that defy belief, enough survived the process of dispersal to ensure the continuation of the species.

Hong Kong's coasts were also richly varied. Two facts explain the marine diversity: the monsoons brought alternating cold and warm sea currents—and the archipelago's eastern waters were predominantly oceanic, while those in the west were estuarine. Thus, besides different *sea* conditions, the two influences also formed very different *coastal* habitats.

Hong Kong's small streams often enter the sea through beach sandbars, such as here at Fan Lau Tung Wan on southern Lantau Island.

Hong Kong's coasts experience both oceanic and estuarine influences. In the east, in places such as here on Basalt Island, ocean waters surround the land.

Four general coastal environments existed here around 5000 BC, as they do today. In the north-east, silt from local streams had created some sand and mud shores, but the streams were too small to lower significantly the offshore ocean water's salinity. The coasts to the south-east, especially on the islands, were exposed to the full force of the ocean. Hence they were steep and rocky—and, except for isolated coves and bays, largely inaccessible. In the south-west the rocky shores were less inhospitable: there, river sediments had created beaches and tombolas (sand spits joining nearby islands), and also made the water much less saline than further east. As for the north-west, countless monsoon loads of Pearl River silt had left a smooth-edged coast that was often more mud-and-mangrove fringe than visible coastline.

Varied marine habitats resulted from this typically Hong Kong interplay of influences. Hence, along an unusually long and indented coastline for so small an area, there was great diversity. Tropical species dominated over temperate ones; and, though rich in species, the local seas were relatively sparsely populated.

Along exposed rocky shores, just above and below the tideline, were barnacles, crabs, limpets, shellfish, and anemones. Further down, mostly below the lowest tides, were other molluscs, seaweeds, and sea urchins. The sheltered sandy shores were equally varied. Burrowing or resting, only appearing with certain tides, lived periwinkles and fiddler crabs; and, lower down, other crabs, shellfish, sea urchins, and starfish. Washed onto the beaches was evidence of the offshore marine life: seaweeds, jellyfish, cuttlefish, starfish—and, of course, countless kinds of fish.

It was to these coasts that people first came. *Homo erectus* was living in Asia at least one million years ago, and the remains of early *Homo sapiens*—some 140,000 years old—have been unearthed in Guangdong.

Homo sapiens was larger than all but a few of Hong Kong's fauna. But, weight for weight, these relative newcomers were puny compared to almost any Hong Kong land animal. However, the species had some telling advantages. They could move on two legs, leaving their upper limbs free; their extremities were remarkably manipulative; and, though virtually blind compared to keen-eyed birds, *Homo sapiens* had reasonable binocular vision.

Above all, in place of other animals' largely instinctive behaviour, *Homo sapiens* observed, evaluated, remembered, and planned. Language allowed thoughts and ideas to be shared. Time was not an endless continuum, but something that could be measured—in suns, moons, and recurring seasons. And, allied to this practical intelligence, there was often a sense of wonder.

Thus equipped, using rocks, plants, or animals—almost anything suited some purpose—men and women gradually mastered their surroundings. By modifying the natural world they inherited, they made life more bearable. Fires could be kindled against the darkness, skins worn against the winter cold, shelters erected against the summer rain. At night the forest need no longer bring terror—and the seasons need not totally dominate life.

About 15,000 years ago, people were living across wide expanses of China and through most of South Asia. By about 8000 BC many of them had relatively advanced skills, and the benefits of turning wild country into farming settlements were becoming increasingly obvious—and achievable. The first Huang He (Yellow River) villages were established about 5000 BC. The entire world population of *Homo sapiens* then numbered about five million—less than the population of Hong Kong today.

So far as is known, humans did not reach Hong Kong until some time after 5000 BC. By then, the seasons were long-established. There were two monsoons and four seasons: autumn, winter, spring, and the long summer. Unusual seasons sometimes disrupted nature's cycles. But, year to year, generally recurring rhythms patterned existence for Hong Kong's plants and animals.

The summer over, the winds shifted to the north. The change brought clearer skies, less rainfall and humidity, and—most importantly—cooler temperatures. Deciduous trees began turning, insects became scarce, and insectivores prepared for their hungry months. Snakes, lizards, and other reptiles prepared to hibernate. Northern birds, fleeing even colder latitudes, began arriving or stopping off on longer migrations. With the summer heat gone, seaweeds reappeared and spread around the coast.

Winters were cool—and occasionally bitter. There was almost no rain. Plants lost their rich greens and forest soils dried until the leaf litter crunched underfoot. Winter fruits ripened with showy colours: purples, blues, and reds. Temperate mammals, evolved for long gestations and summer births, began to pair and mate. Algae spread along the coasts, and visiting seabirds fed off fish and marine invertebrates.

With spring the winds shifted southwards. Under dank, unpredictable skies fog and mists hid the higher hills, and light rains freshened the land. Deciduous trees sprouted red-green leaves; multiplying insect larvae fed off the new foliage. Lizards and snakes, revived by the warmer weather, sunned themselves on rocks. Local birds began breeding, and visitor and migrant birds winged northwards. In sudden bursts, flowering trees and shrubs splashed the hills with colour.

So began the longest season: the hot, humid, rainy summer. Cumulus clouds, brilliantly coloured at dawn and dusk, swept in from the south. Often falling just before dawn, soaking rains left water everywhere: in the slimy ground, in full waterholes, in cascading waterfalls. Beneath the forest canopy, lianas flowered and butterflies flitted and hovered. Revelling in the damp, cicadas drowned out other sounds. Flying termites, after nights of intense activity, shed their wings, crawled off, and formed new colonies or died. Birds built nests, mated, and hatched their eggs. Young mammals, slowly finding their legs, fattened on the summer fruits. And the coast grew drab, as low tides exposed winter seaweeds to the scorching sun.

The southern monsoon also brought another dimension: a season within a season—a few months during the second half of summer when the elements sometimes descended in raging fury. Then ...

For days there was often a thick haze. The darker grey-green of nearby hills merged into the paler grey-green distance, which merged into a formless horizon. Dragonflies hovered and darted, some mindless instinct welcoming the approaching rain. The skies, occasionally almost yellowish grey, took on a solid stillness. The sea turned strangely glassy. At last, after days of waiting, wind tremors shook the leaves, and branches stirred with passing flurries. Animals, their senses aroused, went to ground, and squalls freshened beneath massing clouds.

The storms sometimes lasted hours, sometimes days. At their worst, the expanse of sea was hidden under flying spume and raging swells. Rain fell in torrents, driven sideways, cutting into the earth. Awesome gusts lashed the land, trees bent double. Roots deep in the ground strained against elemental forces.

One of the waterfalls in the Ng Tung Chai gorge, cascading after summer rain.

REGION

THE CENTRAL MOUNTAINS

I gaze ahead at the towering mountain steps,
I watch the morning clouds swathed round the range,
I listen to the evening torrent rushing through the gorge.
Rocks lie strewn here and there like chess-stones,
I scramble over them, marvelling at their strangeness.
This is no mountain, this is no hill
But a very tower, a gate turret....

HSIEH LING-YÜN, about AD 400

It is summer. Beyond a steep ravine one of the Ng Tung Chai waterfalls plunges into a boulder-strewn pool. Cutting through a mountain cleft the cascade plummets past rocks green with slime. On either side sponge-like mosses drip with water that seeps from the hillside. Growing from jagged crevices, ferns sway heavy with water.

Far above, hidden behind the precipice, lies Tai Mo Shan. Below, veiled in drifting drizzle, gurgling through dense thickets, is the lower stream—the source waters of the Lam Tsuen River. How long, I wonder, has it taken for water and rocky debris to carve out the waterfall's 15-metre drop?

In the shaded ravine the air is dank. But Tai Mo Shan's summit, our destination, is still some 700 metres higher—above some of the steepest slopes in Hong Kong. It promises to be a grinding climb. For every 200 metres up, the temperature will drop just one degree.

Halfway to the summit, 300 metres above the waterfall, and 500 above the Lam Tsuen valley, we rest on a saddle—catching our breath, sodden with sweat. We are above the tree line now. Below, the ravine cuts through the wooded mountainside; above, all is grassland except where forest reaches up along the watercourse. The slopes are criss-crossed with old stone terraces, like tribal scars across the grassy chest of Tai Mo Shan.

Later, still 150 metres below the summit, we pause at the source of the waterfalls: a narrow stream running between some mossy rocks, past a grove of stunted trees. Water skaters skim across a pool, and freshwater shrimps hide under its rocks.

Tai Mo Shan is Hong Kong's highest point. The mountain—its name means 'Big Hat Mountain'—is only modestly striking close up. But seen from a distance, as from across the harbour, Tai Mo Shan towers 957 metres over the surrounding coast. Its sheer bulk cannot fail to impress.

Tai Mo Shan forms the centre of a sprawling geological massif. Its foothills reach to the Tai Po Kau valley, about five kilometres north-east, and to the Tai Lam valley, about 10 kilometres south-west. Other wider, flatter valleys separate the Tai Mo Shan complex from ranges to the north, east, and west.

Three Country Parks cover much of this massif: Tai Mo Shan and Shing Mun Country Parks, both about 1,400 hectares; and Tai Lam Country Park, with an area of 5,330 hectares. In addition to this there is the 460-hectare Tai Po Kau Nature Reserve. Although not far from densely populated urban areas, these Country Parks are rugged and virtually uninhabited. Their total area exceeds that of Hong Kong Island by almost 1,000 hectares.

This central region gives the clearest impression of how Hong Kong appeared before man. At Tai Po Kau, one can explore among woodlands similar to the primordial forests. On Grassy Hill and Tai Mo Shan one sees how people later turned the forests to grassland. And in Tai Lam valley one witnesses the stark evidence of how grassland easily degenerates into barren badlands.

Tai Po Kau Nature Reserve lies roughly halfway between Sha Tin and Tai Po. A leafy delight for walkers, for ecologists Tai Po Kau provides a case study of Hong Kong's changing vegetation. Here, without the gymnastics of scrambling into remote gullies and ravines, one has easy access to Hong Kong's finest woodlands. Barely a kilometre up from the old Sha Tin to Tai Po road the noise

This mixed woodland, in Tai Po Kau Nature Reserve, is similar to Hong Kong's primeval forests.

Reforestation helps even run-off after rainfall. Here, days after the last rain, these slopes below Tai Mo Shan still seep water.

of traffic fades. Climb higher and—with a little imagination—one is transported to the Hong Kong of about 5,000 BC, before humans arrived.

However, dense and diverse as much of Tai Po Kau's woodland now is, none of it is primary (or original) forest. For many centuries the trees here, like all of Hong Kong's, were cut by villagers. The trees seen here today were almost all planted by Agriculture and Fisheries Department foresters since the 1960s.

The most recent plantations, those established in the 1970s or after, are still very young. Typically they have a single, dominant species and a relatively undeveloped under-storey. The result is unnaturally uniform habitats and food supplies, and hence fauna that are less varied and numerous than in mixed woodlands. Brisbane Box is the most common plantation species, a fast-growing eucalyptus from eastern Australia. In parts of Tai Po Kau these clean-stemmed trees soar 20 metres or more.

But go deeper into Tai Po Kau, into its higher, steeper valleys, and the plantations give way to mature woodland. Native South China trees, many self-propagated in the composts provided by plantations, are mixed with planted species. In the more inaccessible parts, there are almost certainly pockets of pre-war trees and remnants of much older village woods—with Golden Bamboos, Chinese Banyans, Camphor Trees, Incense Trees, Lacquer Trees, Tallow Trees, and Chinese Fan Palms. The species mix may have changed, and the trees are younger, but here one comes close to primordial Hong Kong.

I pause along a red-earth track walled with mosses. Crystal springs flow from the hillsides. Light and shade, dull greens and vivid emeralds, pattern the forest. Lichen-spotted boulders lie tumbled on fern slopes. Trees, some with developed buttresses, climb to a canopy that blocks out the light. Moisture trickles down leaf stems, along branches, onto the damp, decaying leaf litter. A stream splashes past rounded stones, dragonflies flit and hover above a translucent pool.

Over past centuries, as Hong Kong's forests were cut back, the larger animals were hunted out and the forest birds replaced by grassland and shrubland species. But today, as woodland habitats expand with the reforestation, Tai Po Kau's fauna bears some resemblance to the animals of ancient Hong Kong.

Twenty years ago virtually the only forest birds seen in Tai Po Kau were winter visitors. But after that, as the older plantations matured into diverse woodlands, forest birds became regular visitors—and, over the past two decades, some have returned as resident species and appear to be increasing. These South China birds, rarely seen elsewhere in Hong Kong today, may well have lived in the local primordial forests—among them Great Barbets, Scarlet and Grey-throated Minivets, and Emerald Doves.

Some of Hong Kong's larger mammals also appear to be recolonizing parts of Tai Po Kau. These animals, timid and often nocturnal, mostly keep to the thickest woodland, so the only definite evidence of their presence comes from droppings, occasional sightings, and remote-control photographs.

Since the mature woodlands are similar to the primeval forests, since development elsewhere has forced animals out of remnant forest habitats, and since illegal hunting has declined, one might expect small increases in these woodland populations. Indeed, although still rare, sightings have increased: of Barking Deer, Chinese Pangolins, Ferret Badgers, Masked Palm Civets, and Small Indian Civets. Similar increases are also occurring at another woodland refuge, Kadoorie Farm and Botanic Garden, just north of Tai Mo Shan.

Rhesus Macaques, Hong Kong's only indigenous monkeys, had been virtually (perhaps completely) wiped out through tree-cutting and hunting by about 1900. Those seen today in Tai Po Kau and in the Kowloon hills are almost all (perhaps all) descended from macaques since released by man. Today the forest-dwelling macaques, grey shapes often seen swinging through the trees, are lithe and aggressive. But in Kam Shan Country Park, on the fringes of Kowloon, the tribes that loiter there have been made obese and squalid by humans.

Reptiles are widespread throughout Tai Po Kau. Besides various harmless snakes, there are hooded King Cobras which prey on other snakes. Although well camouflaged, sometimes seen among leaf litter or rocks, are lizards and skinks such as the russet coloured Brown Forest Skink, and the stumpy, reddish bellied Chinese Skink. In summer female Woodland Spiders, large and

A stream runs through a glen, in Tai Po Kau Nature Reserve.

Hong Kong's lowlands are now virtually all 'developed'. Here, seen from Lead Mine Pass, the lights of Tai Po are etched against the rugged Pat Sin range.

Grassy Hill, eerily swathed by gusting, early morning clouds and mist.

yellow-black, create striking circular webs, often a metre or more across, where they capture insects on the wing.

It is late autumn. The weather is far cooler and drier than three months before, when I climbed past the Ng Tung Chai waterfalls. After a day hiking up through Tai Po Kau, I camp at Lead Mine Pass above the Shing Mun valley. To the west the massive bulk of Tai Mo Shan rises up, its capping of cloud framed in the twilight; to the east, Grassy Hill blocks out Sha Tin; and to the north, far below, Tai Po's myriad lights are enclosed at dusk by blue hills. Insects are the only sound.

Bedded down out of the wind among some grasses, I sleep well. About five that morning I begin climbing Grassy Hill, aiming to be there for dawn. Pushing between chest-high grasses—the summer's lush growth—I trudge up through the darkness. There has been a heavy dew and the clayey track is treacherous. Jabbing my tripod into the ground, I move slowly up.

Low clouds are drifting in. The last beckoning star, then the higher hillside, finally even Tai Po's lights, are lost in misty vapours. Stands of dead trees, stark and jagged, make the scene eerie. Further up they give way to grassy boulder slopes. The last sharp rise is capped with a rocky knoll, then—the summit of Grassy Hill, 647 metres. It is some 250 metres above Lead Mine Pass, and slightly higher than Victoria Peak on Hong Kong Island.

Somewhere to the east, out across Mirs Bay, the sun is rising—but on Grassy Hill dawn is grey and windswept. Scudding clouds give hints of faint blues and pinks, tantalizing colours that quickly fade. An hour after sunrise, the cloud and mist thin—and the sun momentarily bursts through. Mesmerized, huddled in a space blanket against the damp cold, I fumble with my cameras and lenses. As the pioneer conservationist Henry David Thoreau, camped on Ktaadn, a New England mountain 'deep within the hostile ranks of clouds', wrote in 1846:

> *Now the wind would blow me out a yard of clear sunlight, wherein I stood; then a gray, dawning light was all it could accomplish, the cloud-line ever rising and falling with the wind's intensity.*

Soon afterwards, a rhythmic, metal tapping echoes up through the valley above workaday Sha Tin. But where I am, still swathed in mist, the only realities are grass and rock and wind—and momentary shafts of brilliant light.

Three hours after dawn the clouds suddenly lift. A panorama of staggering variety opens before me: the central New Territories' hills, valleys, and communities—with their story of massive environmental change. Below the summit, in place of primordial forests, grassy slopes roll down to Lead Mine Pass. Beyond there reforested hillsides reach into the deeper valleys. And around Tolo Harbour concrete towers march up the lower slopes.

From Grassy Hill centuries-old stone terraces can be seen criss-crossing the surrounding slopes. Daunting monuments to toil, the terraces were built at least 250 years ago, and were almost certainly used for growing tea. The remains of the higher ones, usually only visible as shadowy shapes beneath tall grasses, are clearly visible after fires. Those further downhill, however, are now always hidden among the reforestation. Remarkably little is known about these terraces, which can be seen today on hillsides across much of the central New Territories. Sloping diagonally, they were intended for crops such as tea needing well-drained soil.

Grassy Hill also reveals a very different era: modern Hong Kong, with its massive industrialization and urbanization. Deep in Shing Mun (Castle Gate) valley is the Shing Mun Country Park, where the Shing Mun (or Jubilee) Reservoir nestles among Australian Paper-barks. The dam's blue-green waters point across the hills to grey, industrial Kwai Chung. In the opposite direction, tower blocks line Tolo Harbour near Sha Tin and Tai Po. And across Tolo Harbour is Plover Cove Reservoir—a dam so vast that it makes Shing Mun Reservoir, built only thirty years earlier, seem almost toy-like.

Infinitely slowly, streams descending from Tai Mo Shan carved out this valley.
The view looks north, over the Shing Mun (Jubilee) Reservoir to Lead Mine Pass. Grassy Hill is on the right.

Hong Kong's mountain streams, like these headwaters on Tai Mo Shan, remain crystal clear; but the lowland streams are all polluted.

Later that day, after dropping down to Lead Mine Pass, I begin climbing up to Tai Mo Shan.

Hong Kong's mountain paths mostly follow old village tracks, such as this ridge-top route—part of the old way from Tolo Harbour to the Pearl River estuary. The boulders along the track, I guess, are barely changed since people first came by, like the boulders on Thoreau's Ktaadn:

> *A vast aggregation of loose rock, nowhere fairly at rest ... the raw materials of the planet dropped from an unseen quarry, which the vast chemistry of nature would anon work down into the smiling and verdant plains and valleys of earth.*

Tai Mo Shan's 957-metre summit stands high above the New Territories, clearly visible from as far away as the Shenzhen River. Indeed, as a Guangdong gazetteer recorded in 1688, it was always one of the most notable features of the local landscape: 'It is a big mountain in the Fifth Division, with a stone pagoda and many tea plantations.'

About a kilometre to the north stands Kwun Yam Shan (Goddess of Mercy Mountain) in Kadoorie Farm. A near-vertical peak, Kwun Yam Shan is so honeycombed with fractures that warm valley air actually rises through the formation to its summit. Often swathed in mists, Kwun Yam Shan has been regarded as sacred for centuries; near its summit are ancient stone altars.

Tai Mo Shan, Hong Kong's topographical hub, is also its climatic centre—as the clouds that often hang over the mountain suggest. This is the wettest part of the Territory. The summit's annual mean rainfall of 3,000 mm is well above the Central and Kowloon figure of about 2,000 mm, and twice that for Waglan Island off Shek O.

Hardly surprisingly, Tai Mo Shan is also one of the windiest places in Hong Kong—and the coldest. Frosts settled there during the historic cold spell of January 1893. In December 1991, when bitter winds surged down from Siberia, Hong Kong shivered. Near Tai Mo Shan's summit the temperature dropped to –5° Celsius, and drizzle spangled the leaves with ice.*

The cool, wet conditions on Tai Mo Shan's northern slopes nurture some rare plants. Grantham's Camellia was first discovered there in 1955—and other camellias, including the Hong Kong Camellia, are common. Village fuel gatherers no doubt harvested Tai Mo Shan's shrubs and trees. However, hidden among its most inaccessible gullies, there may be trees descended from the first forests, since some species are found there but nowhere else.

Hong Kong's hills are a maze of watercourses. But, excluding the Shenzhen River, no other stream can be truly called a river.

The Lam Tsuen and Shing Mun rivers, the largest Tai Mo Shan watercourses, are 'rivers' in name only. Their source streams tumble downhill through ravines and over waterfalls, but their valley courses were never wide. Today, near where they enter Tolo Harbour, both are simply massive concrete gutters—degraded and polluted.

Remarkably, the upper courses of both streams still support varied fauna, adapted either to the rapid headwaters or to the intermediate pools. Besides tiny water skaters and numerous other insects, there are reptiles, amphibians, freshwater fish, shrimps, and crabs. Chinese Big-headed Terrapins and Three-banded Box Terrapins inhabit the upper pools (though the latter, valued for 'medicinal' uses, are increasingly rare). Among various frogs and toads, the Hong Kong Cascade Frog, unique to the Territory, has toe suckers that allow it to cling to rocks despite fast currents. Minnows live in the higher pools; and two small fish species, elongated and large-finned, swim against the turbulent source waters and graze on underwater algae.

Hong Kong's streams are highly seasonal. They cascade after intense summer rainstorms, flow steadily through normally wet summers, and subside to mere trickles, and often only pools,

* *Despite the cold, few plants near Tai Mo Shan's summit were killed, and some actually flowered soon after. This indicates the adaptation of Hong Kong's native flora to the extremes of the tropical and temperate seasons.*

The Shing Mun valley was mostly treeless in the 1960s. Rich woodlands now spread across its lower parts, and, as here, up its middle slopes towards Tai Mo Shan.

Eroded granite badlands, stripped of their topsoil, at Tai Lam Chung Reservoir. Plantations have been established on the surrounding, less eroded slopes.

during the dry winters. Around Tai Mo Shan, as throughout the Territory, old granite slab bridges testify to the force of summer torrents. Indeed, without them the village pathways would have been often impassable. As late as 1950, Tai Mo Shan's summer flow was still being used to drive eight mills that ground Incense Trees into powder for joss. G. S. P. Heywood described one village then, along Tai Mo Shan's southern flanks:

> *A dozen little houses [are] perched on the sides of the valley, each with its waterwheel busily turning.... The atmosphere is thick with fragrant dust, and through it you can dimly see great stone-headed hammers pounding away at the aromatic wood.*

Ten kilometres south-west of Tai Mo Shan's summit is the deep Tai Lam (Big Olive) valley. Tai Lam Country Park, the Territory's second largest, includes the entire valley and its surrounding slopes. Much of the Park has been reforested in recent years, but near the southern end of Tai Lam Chung Reservoir the erosion caused by earlier deforestation is stark.

Unlike Tai Po Kau's and Tai Mo Shan's volcanic country, the Tai Lam valley rests on granite. Hong Kong's volcanic soils are generally richer than the granitic ones. But more critically, given the Territory's precipitous slopes, the two kinds of rock have widely different abilities to withstand weathering and erosion—as Tai Lam dramatically shows.

Volcanic rocks, like Tai Mo Shan's and Tai Po Kau's, weather (or decompose) underground relatively slowly; and so, even when deforested, their surfaces erode only slowly. However, heavily fractured granite, like Tai Lam's, weathers very deeply, especially with Hong Kong's heat and rainfall; and so, when stripped of trees, its surface erodes rapidly—and often disastrously.

Tai Lam's story is an environmental vicious circle. Once Tai Lam's primordial forests had been destroyed, or much reduced, through tree-cutting and fires, the area's weathered subsoils were still more exposed to heat and rain. Downpours scoured away the topsoil, so increasing the area's susceptibility to deep weathering. The loss of topsoil made it harder for plants to take root—which increased the weathering—and so it continued.

The end result was both long-term and short-term land degradation. In the long term (since even before the primordial forests were there) weathering and erosion has left the Tai Lam hills much lower and more rounded than the higher, sharper peaks around Tai Mo Shan and Tai Po Kau. In the short term the loss of trees hastened the erosion of Tai Lam's best soils, those fertilized over millenia by the original forests.*

Hence the cowboy badlands of today: convoluted, deeply dissected terrain, criss-crossed with ridges and gullies, covered with sterile gravelly soil, and awash with yellow-red water after every rainstorm.

Post-war government reforestation has created plantations in all but Tai Lam's worst parts. However, only some species are able to establish in such poor conditions: Chinese Pines, Slash Pines, Australian Forest Grey Gums, and Lemon-scented Gums among them. Moreover, since about 1982, introduced pinewood nematodes have killed off countless Chinese Pines, especially in the Tai Lam valley. Thus, around the Tai Lam valley, mature woodlands such as Tai Po Kau's have not developed—and perhaps may never do so.

Tai Po Kau and Tai Lam are two extremes. In Tai Po Kau I often felt, as Thoreau wrote, that the woods are 'choke full of honest spirits'—that the forest is 'not an empty chamber, in which chemistry [is] left to work alone, but an inhabited house'. Around Tai Lam the opposite often seems closer to the truth.

It is mid-afternoon, and a friend and I are hiking down into the Tai Lam valley where I hope to photograph the southern badlands. The track around the reservoir winds on and on. Too late we realize that our earlier map reading was hasty: in Tai Lam valley three kilometres as the crow flies means six across the ground, for every few hundred metres a deep gully demands a lengthy detour.

The autumn sun is weakening. An hour and the best afternoon light will be gone, and our hike to photograph the Tai Lam erosion perhaps wasted. We start striding out. Soon we are jogging, our gear bulky and heavy: sweat runs, our backs ache. Fortunately the light holds longer than expected, and around five o'clock we climb onto a pine-clad ridge. It is a moment to savour. There, lit in the late afternoon glow, are the weirdly shaped islands we seek—eroded badlands rising out of the reservoir's waters.

* *As David Dudgeon and Richard Corlett write in their ecology of Hong Kong: 'Tall, closed forest cannot develop on the shallowest soils while any significant increase in soil depth will take hundreds of thousands of years.'*

When people first reached Hong Kong they found mostly subtropical forest, such as here in Tai Po Kau Nature Reserve today.

HISTORY 5000BC – AD1841

SETTLING THE LAND

Since [the farmer of South China] is at the mercy of Nature, and Nature to him includes the wind, the tides, the surrounding hills, the sun, the moon, and the entire universe, he feels that he must propitiate, outmanoeuvre, or come to terms with its significant forces.

ARMANDO DA SILVA, 1972

Throughout China, from time immemorial, there has been an abiding sense of nature's power. Across a vast landscape, where most people barely survived as subsistence peasants, the threat of hunger or even starvation was always present. Unkind seasons could ruin harvests. Cruel weather could destroy crops, houses, and boats—devastate whole regions, and shatter lives.

Hong Kong was no different. Farmers and fishermen there, keenly attuned to their surroundings, sensed the weather's moods. They needed no instruments to predict its changes. Typhoons, however, were an exception: for, though their approach could be generally felt, their direction and strength were beyond knowing. This unpredictability heightened the dread of typhoons, especially among farmers and fishermen.

The *Xin'an Gazetteer*, a record of events in Xin'an County (which included the future Hong Kong), describes the results of typhoons (*dai fung* or 'big winds'). Typhoons, the *Gazetteer* says, usually come in 'the 6th, 7th and 8th' months: 'Still clouds build up thickly, there is continuous rumbling thunder, the sea seethes, the water pounds on rocks with great noise, and sea birds fly panic stricken before the wild gusts.'*

In June 1643, for example, a typhoon 'uprooted trees, demolished houses, and capsized many boats'. In September 1677 a typhoon raged out of the night: 'A large number of people and livestock were crushed to death in the ruins [of a city wall]'. Torrential rain almost always accompanied typhoons, as in October 1818, when 'bridges and embankments were all washed away in rivers of mud, and hill fields collapsed in landslides'.

Sometimes, however, the pattern of heavy summer rain was disrupted. In 1636 there was almost none and the first paddy harvest was lost; only government grain supplies prevented widespread famine. The rain in 1804 was incessant, but it came in February, thus retarding the winter salt production, which relied on sunshine for drying sea salt.

Beginning with the arrival of prehistoric people, and continuing until the establishment of the Colony of Hong Kong, the primordial natural landscape was radically altered by man. But, seen over the whole period, the change was mostly very gradual—for the era spanned some seventy centuries, from about 5000 to 4000 BC until AD 1841.

Neolithic and Bronze Age people were too few to cause significant environmental change, and for some centuries after the Hong Kong area had been absorbed into the Chinese empire the degree of change was slight.

However, significant changes came to Hong Kong's natural landscape with the Tang Dynasty—and even more during, and after, the Song Dynasty. By then a large Chinese population had reached the region, establishing agricultural villages in the accessible lowlands. These settlers cut down the forests, ploughed fields into virgin soil, and exploited the wildlife.

Over the centuries, their cumulative activities caused far-reaching environmental change. This became more widespread when other migrants later moved into the area, settling in the empty uplands and other remote, marginal areas.

This saga of environmental transformation—from primordial nature to an agricultural landscape—resulted in both losses and

* *The* Xin'an *(San On)* Gazetteer *appeared in two editions, in 1688 and 1819. The 1819 gazetteer, as its modern editor Peter Ng notes, 'constitutes the single, large-scale, comprehensive work on the area before the coming of the British'.*

Wild grasses grow on old agricultural lowlands, once rich paddy fields. Agricultural people first settled such fertile low lying areas.

Striped Bamboo, pictured here, was the most common species of bamboo which grew in the Hong Kong forests — and one of the most useful plants to man.

gains. As the population grew, the pre-existing flora and fauna undoubtedly suffered. Some species disappeared. But the settlers also enriched the landscape, and they invested it with human meaning and stories.

The precise origins of the people who first reached Hong Kong are uncertain, but they probably came from around Indo-China. These seafarers, generally known as Yueh, arrived here some time between 5000 and 4000 BC. Typhoons apart, Hong Kong's sheltered bays are more welcoming than its abrupt hills; and not surprisingly the Yueh stayed around the indented coast, harvesting the sea's bounty and using the land only incidentally.*

During the following three millenia more advanced Neolithic cultures slowly developed around Hong Kong, influenced by contacts with China. By about 1200 BC Bronze Age implements—metal knives, axe heads, spear tips, fish hooks—had brought a new mastery over the environment. By then there may have been fixed settlements, and people may have been growing brackish water rice in tidal flats. But, as the Chou Dynasty dawned, it is likely that the Yueh's descendants—far from seriously modifying the natural landscape—were still themselves dominated by it.

* *Until recently 4000 BC seemed the most likely date for the first settlement of Hong Kong. However new archaeological discoveries indicate that people may have arrived here closer to 5000 BC.*

The Chou Dynasty (722–222 BC) saw at least intermittent contacts between the Pearl River region and North China. Then, in 214 BC, during the short-lived Qin Dynasty, imperial forces occupied the Pearl River area and established local garrisons.

From the Han Dynasty (206 BC–AD 220) on, the Pearl River delta, including what later became Hong Kong, was a definite part of China. However, though some Chinese went there and intermarried, there was no significant Han Chinese migration to the region. Settled agriculture had not developed in the Hong Kong area, though moving 'slash-and-burn' farming was probably practised. Changes to the environment, we may thus assume, were still extremely slight—except perhaps immediately around the coast. By late in the Han Dynasty, China had about one quarter of the world's 200 million people, and deforestation in North China was already severe.

Han officials considered South China remote and wild. Indeed, compared to North China's dry loess plains, the region was rugged and daunting. An official banished to South China in 303 BC wrote of 'dark and interminable forests, the habitation of apes and monkeys' with 'mountains wet with rain mists, so high that the sun was hidden'. South China, to Han writers, was truly primordial—a place, one poet wrote, where branches 'twist and snake':

In the deep wood's tangle
Tigers and leopards spring.
Towering and rugged,
The craggy rocks, frowning.
Crooked and interlocked,
The wood's gnarled trees.

The perception of the south as a hostile wilderness had been much softened by the Tang Dynasty (AD 618–906). This reflected, in part, the great southwards shift of China's population that occurred in the chaotic, disordered centuries between the Han and Tang dynasties—when people fled beyond the Yangtze River towards the subtropics. Tea drinking spread across China after this migration, as it brought northerners to the wet south that favours tea growing.

The future Hong Kong became a part of Dongguan County in 757. Guangzhou was by then an important, if peripheral, Chinese trading centre. It was linked by canals and rivers to North China and by sea to as far away as Arabia. During much of the Tang period Hong Kong exported pearls, salt, and lime produced from kiln-fired coral or seashells. How much shrub- and tree-cutting for the lime kilns reduced the local forests can only be guessed at, but it must have led to at least some local deforestation.

The local population, even late in the Tang Dynasty, was still very sparse. There was little intensive farming, though the hills were probably fired to expel wild animals and produce ash fertilizer; and around the coast boat people fished and grew rice. Because of the silence concerning them in Chinese records, details about these illiterate people are shrouded in mystery. For instance, an official report made a few centuries after the Tang Dynasty, quoted by the Hong Kong scholar Lo Hsiang-lin, tells us virtually nothing about the lives of Hong Kong's pearl divers—but a good deal about its Chinese author's haughty superiority:

> *The ruggedness of the landscape and the savageness of the inhabitants of the pearl fishing areas render it unwise for the local government to push its authority ... [The pearl divers] look more like beasts than men, go naked, and have tongues resembling those of birds, and are capable of diving like otters.*

Moreover, reports such as this say nothing about the degree of environmental change. Long-nosed Crocodiles had certainly once inhabited South China's river estuaries, as the archaeological record shows—and as the carved heads on some dragon boats suggest. But we have no knowledge of when these magnificent creatures were hunted out of existence around Hong Kong.

An explosion of technical knowledge marked Song Dynasty (960–1279) China, then the world's most advanced society. New methods of reclaiming, irrigating, and fertilizing the land greatly improved agriculture; and new strains of grain, especially Indo-Chinese rice that could be double-cropped, greatly improved yields. Less happily, increased industrial demand for charcoal and timber led to severe deforestation in North China.

Meanwhile, population growth in North and Central China created an urgent need for new land to settle. China's population still numbered only about fifty million near the beginning of the Song Dynasty, but over the next 200 years it more than doubled. This population growth, aggravated by the southward march of the marauding Mongols, displaced countless millions of people from China's heartlands.

These terraces, near Double Haven, highlight the toil of farming Hong Kong's marginal country, and the deforestation that village life inevitably brought.

Striking rock formations, such as this granite tor on Lamma Island, were often considered propitious or spiritually significant by villagers.

Most migrated southwards. During the later Song Dynasty, the Yuan (Mongol) Dynasty (1280–1367), and the early Ming Dynasty (1368–1644), thousands of people reached Hong Kong's fertile lowlands. Most were humble peasants, but some had education and wealth. Not for the last time, émigrés fleeing disorder in China at last saw Tai Mo Shan rising majestically beyond the Shenzhen River. Nor was it the last time that waves of newcomers—rebuilding their lives in this subtropical haven—inevitably and radically transformed its natural landscape.*

Well before the sixteenth century Hong Kong's best farming land, the alluvial plains between the Shenzhen River and Tai Mo Shan, had been closely settled by the Punti—as these people later came to be known. The name means 'this place', and was used later to distinguish the Punti from subsequent immigrants; for clarity, the word is used here throughout.

* *In what follows, 'Hong Kong' refers to the land now comprising the Territory: the New Territories, Kowloon, Lantau, Hong Kong Island and the other islands. Previously part of Dongguan County, Guangdong Province, from 1573 the area of the future Hong Kong was included in Xin'an County, as remained the case until 1898. The county town was Nantou.*

The Punti were skilled sedentary agriculturalists for whom Hong Kong's forested hills and valleys represented potential income and, with good fortune, wealth: land to be cleared, land to be tilled, land to be passed down the generations. Across fertile valleys they located fields, huts, and houses, and later built ancestral halls and temples. Clan allegiances and traditions gradually patterned the land, and the language of the Punti—Cantonese—took root.

Meanwhile the Tanka, local boat people perhaps descended from the Yueh, continued making their living around the coasts. Now outnumbered and often scorned, they faded into Hong Kong's sea-hazed background on their weather-beaten junks and sampans.

Two inter-related issues guided the Punti in deciding exactly where to settle: the natural landscape's physical potential and its spiritual qualities. The former—soils and streams, slopes and shelter—were obvious enough to skilled farmers. The latter—the

Chinese Banyans, a species of fig, have few practical uses but live to a great age. Partly as a result, banyans growing near villages were revered and often had shrines.

The paths that today cross the Hong Kong uplands, such as this one near Ma On Shan, were once byways linking mountain villages and the lowlands.

complexities of *fung shui*—demanded much investigation. Settling on land with poor *fung shui* would have seemed, to the Punti, like inhabiting a body with brittle bones and a weak heart.

Fung shui (literally 'wind water', and used here with the Cantonese spelling, rather than the pinyin *feng shui*), the Chinese practice of geomancy, is especially prevalent in South China. Among other things, it influences the siting of virtually any man-made feature: fields, houses, temples, and above all, graves. More than mere superstition, *fung shui* is perhaps best described as a belief in the spiritual power of natural elements and forces. Good *fung shui* sites are in harmony with these forces, harnessing their energies or *qi*, thus minimizing human discomfort or suffering.

But *fung shui* also reflects purely environmental aspects that most people would prefer anyway. Take, for example, the views of one Han Dynasty Taoist recluse who wrote:

> *Let my dwelling have good fields and a spacious homestead, with its back to mountains and looking on a stream, with channels and pools surrounding it, bamboos and trees set on every side, a threshing-floor and vegetable garden made before, and an orchard planted behind.*

Remote as these scholarly reflections were from the hardships of peasant life, the ideal was in fact very similar to the auspicious *fung shui* for any South Chinese village: hills surrounding the buildings, a stream winding down past gentle slopes, trees (especially bamboos) planted behind, some fields in front, and, ideally, a view over water or the sea. *Fung shui* may thus be seen as a 'rural landscape language', important to illiterate peasants as a way of ordering their environment. Recent research indicates that the prevalence of 'wet rice' growing in South China may be a result of *fung shui* beliefs, since paddies are situated on the lowlands with villages and graveyards above.

There was also a much broader aspect to *fung shui*, reflecting both regional topography and indeed the features of the entire Chinese land mass. And here Hong Kong was—and is—most fortunate in its uplands.

Mountains, Chinese belief says, are the abode of immortal gods and dragons. Relationships between nearby mountains and hills control the balance of *yin* (tiger) and *yang* (dragon) energy, sites with ideal *fung shui* being those in which dragon and tiger energies flow together harmoniously. The forces that permeate Hong Kong's hills originate (so *fung shui* experts believe) in the Kunlun Mountains bordering Tibet—the home of China's most powerful dragon (and the source of its great rivers). From there, mountains, the veins along which *qi* flow, extend across China and finally reach Hong Kong. These veins, it is said, dive below Victoria Harbour to end their journey on Hong Kong Island.

The Punti clans that settled around Kam Tin, in the north-west New Territories, found excellent *fung shui*. Kwun Yam Shan rose nearby, Tai Mo Shan towered over the entire area, a dragon and tiger lay on either side. Moreover, Kam Tin's alluvium was as fertile as any local soil. Was it any surprise that Kam Tin grew some of Hong Kong's finest rice, or that its clans prospered? Or that, in recognition of their generous disaster relief in 1587, a Xin'an County official reputedly gave Kam Tin its name: Golden Fields?

The evidence concerning the Punti's and Tanka's impact on the natural landscape is very scanty. Nor is it known how much Hong Kong's natural setting had been altered already, though when the Punti arrived much of Hong Kong—especially the uplands—was almost certainly still covered with subtropical or tropical broad-leaved forest. What follows, therefore, can only be a general impression, not a quantifiable analysis, of subsequent changes.

When arable land became scarce, farmers cut back the mangroves surrounding parts of the coast and reclaimed the intertidal mud-flats for growing brackish water rice. Mangrove trees have a high tannin content, and the Tanka also cut them to use in preserving their natural-fibre nets and sails from salt water. The scene that a European visitor witnessed at Deep Bay in the 1870s had probably changed little in centuries:

> *A colony of fishers [were] boiling their nets in an exceedingly tall vat, containing a decoction of mangrove-bark, which produces much the same rich brown colour as our fishers extract from alderbark. Here, however, it is considered necessary to subsequently steep the nets in pig's blood to fix the colour.*

Besides cutting mangroves, and of course using timber for boat building, the Tanka's other significant impact on the land ecology lay in the large amounts of grass and brushwood they used to 'bream' their boats. Breaming, the process of regularly burning and scraping boats' hulls, was essential to remove marine organisms and borers. Junks and sampans careened on remote beaches at low tide, with flames licking up from bundles of grass, were a common sight in Hong Kong as late as the 1960s.

The amount of grass, shrubs, and trees the Tanka cut pales into insignificance beside the Punti's inroads into the local vegetation.

Vast areas of China had been deforested long before by the sheer power of peasant numbers. Knowledge and laws concerning forest conservation existed in China as early as 300 BC, and the planting and preservation of forests in specific areas was encouraged. But little stood in the way of widespread tree felling to clear land for farming, as happened almost everywhere around the world where agriculture was established. China's high population demanded, and drove, the deforestation—as 'Clearing the Fields' in *The Book of Songs* suggests:

> *We clear the grasses and trees,*
> *We plough and carve the land,*
> *Two thousand men and women scrabbling weeds*
> *Along the low wet lands, along the dyke walls....*

In Hong Kong the Punti's reasons for tree-cutting no doubt seemed convincing to them. Indeed, they were mostly essential: to clear land for crops and grazing; to make tools, carts, houses, and furniture; to clear paths; and to destroy the lairs of dangerous animals. The Punti also regularly set fire to hillsides. Some fires were accidental, often caused by burning offerings at the annual grave festivals. But most hill fires were deliberately lit: to reduce undergrowth, to flush out pests, and to fertilize lower slopes with ash.*

Probably most damaging of all—year in, year out—people cut trees, shrubs, and grass for fuel. Over the generations prodigious quantities of fuel, and many millions of trees, must have been consumed in Punti and Tanka hearths and braziers.

No single decade, or even generation, probably revealed how drastically the land was being deforested. But over time tree-cutting and burning clearly led to a great loss of forest cover. Long term droughts, by encouraging fires even in normally lush areas, may have speeded the deforestation. Allied with this came more rapid erosion of the topsoils, especially on steeper hills or where the secondary growth was soon cut back. Landslides became more common. The increase in erosion was, over the longer term,

* *Père Armand David, a Basque missionary and naturalist, commented in 1875 on the Chinese 'passion for ceaselessly burning down mountains'. Obviously then in a still-forested region, he wrote: 'With the great trees will disappear ... all the animals, large and small, which need the forest.*

severe. And the soils lost—the primordial topsoils formed over tens of thousands of years—were virtually irreplaceable.*

Except in its inaccessible uplands, Hong Kong thus lost almost all its woodland habitats. As a result, the larger forest mammals either moved away or were hunted out. Only the smaller mammals survived, by retreating into remnant woods, such as those tucked away in the numerous mountain gorges. The forest birds gradually went elsewhere, while grassland and shrubland species took their place.

The change from a natural to an agricultural landscape brought grasslands to the hills and fields to the valleys. The transformation involved massive human effort, helped only sometimes by draught animals. The Punti had specific phrases for the toil of settling the land, as the Hong Kong rural historian David Faure has recorded. *Lok-taam*, 'putting down the carrying poles', meant the period of journeying to new lands was over, and so a settled life must begin. Work must start on *maoliao*, 'wooden sheds' to live in until a house could be built. Then came the long, laborious *hok-kei*, 'opening the fields'. And so the vocabulary continued: building terraces, reclaiming land, digging wells....

Hong Kong has only short, rapid streams, no major rivers, and not a single lake. To the Punti this was challenging and, if the summer monsoons failed, critical—for their two-crop paddy cycle needed water for many months.

The paddy fields that spread across the lowlands demanded prodigious effort. Constructing their irrigation systems was extremely laborious, as was growing their 'wet' rice—from planting out seedlings through to the harvesting. A steady amount of water moving through the interconnected fields was essential, as algae and manures carried in the water gave the fast growth that allowed two summer crops. Diverting streams and forming irrigation channels were thus fundamental to the Punti's livelihood. That Hong Kong's only modestly fertile soils produced some of China's best rice indicated the Punti's industry and ingenuity.

Rich manure smells permeated every rural village. Wooden buckets or pottery urns were strategically placed, near houses and in the fields, for people to urinate into. Villages generally had communal latrines, built near incinerators so people could scatter ash over their waste. Children collected the droppings from buffalo, pigs, and poultry—which, as livestock were scarce, was invaluable. The excrement and urine were kept separately for some time to 'improve', then later used in various fertilizers.

Despite their widespread deforestation, the Punti husbanded the land in other ways. On *chung shan*, or 'pine hills', they grew Chinese Pines to harvest for firewood and timber. And near villages they established *fung shui* woods. These groves, usually established behind (and ideally above) villages, included both planted trees and remnants of the original forest cover. Well-sited *fung shui* woods gave shelter from winter winds, shade from summer skies, and safety from landslides. They also enhanced the local *qi*, so helping make the village auspicious.

Fung shui woods were also store-houses of income—and, if only incidentally, of botanical species. Bamboos, Cotton Trees, Camphor Trees, Incense Trees, Lacquer Trees, Tallow Trees, Chinese Fan Palms, and many other useful species were planted. These woodlands held huge numbers of species; and in Hong Kong *fung shui* woods grew almost half the herbs and other plants known to Chinese herbalists. Though some woodlands were harvested, their trees were never actually cut; and in some villages spiritual taboos prevented access to the local *fung shui* wood. No *fung shui* wood was complete without at least one venerable Chinese Banyan, its base usually holding an earth-god shrine.* The *Xin'an Gazetteer* notes of the Chinese Banyan:

> *Its wood is gnarled and so cannot be used as timber, it will not flame so cannot be used for firewood. Its very lack of useful properties ensures is invulnerability and survival.*

Villages, especially the poorer ones, were often squalid. Sickness and disease were common: infant mortality was high and life expectancy low. Throughout Hong Kong malaria, dengue fever, smallpox, and leprosy were endemic.

The settled lives of Hong Kong's people were torn apart in 1662. That year Qing Dynasty (1644–1911) authorities, fearful that South China's coastal villagers were still loyal to the Ming rulers, ordered the complete evacuation of a strip along the coast 50 *li* (about 25 kilometres) wide. European traders had been in the area since the mid-sixteenth century. Their activities, and increasing piracy, may also have prompted the 'evacuation'.

Around Hong Kong, especially on Lantau and the other islands, some people defied the exile and remained. However almost all the mainland New Territories villages were emptied, and their people

* *The formation of Hong Kong's primordial soils probably took at least 100,000 years—and possibly much longer. The extremely long period makes them, effectively, a 'non-renewable' resource.*

* *Richard Webb, who recently completed a detailed social–botanical survey of Hong Kong's remaining* fung shui *woods, helped with this. These local woodlands, he notes, fit a much broader picture of 'spiritual woodlands' found throughout the world.*

By 1841 Hong Kong's primeval forests had been reduced to mere remnants, mostly in gullies and ravines. Grassland and shrubland dominated the country, as here near Sai Kung today.

summarily forced to trudge inland. Seven years later, in 1669, the order was rescinded. But by then there was a bitter legacy: of hardship, hunger, starvation, and death. As the *Xin'an Gazetteer* recorded after the evacuation:

> *The authorities treated the people as no more than ants and made no provision for relief; relatives treated each other as mere dirt and did not observe the niceties of dutiful behaviour.*

Around Hong Kong abandoned fields and villages had become derelict. Fast growing summer grasses had covered hard-won fields and terraces; irrigation channels had clogged with silt and weeds; creepers had invaded houses and temples; livestock had wandered off; and wild animals had probably increased. 'Large numbers of tigers caused many casualties in 1680', the *Gazetteer* reports.

The *Gazetteer's* population figures speak powerfully: in 1642, well before the evacuation, the population of Xin'an County was 17,871; during the evacuation it fell to 2,172; and in 1672, three years after the order was reversed, it was still only 3,972. Disastrous epidemics between 1639 and 1644 probably accounted for part of the population loss.

Some of Hong Kong's Punti people never returned: many had died in exile. In their place, tempted by land left vacant, came another wave of land-hungry peasants: the Hakka, or 'guest people' as the Punti called them. Originally from Fujian and Jiangxi provinces, distinct in dress and speech, the Hakka were to remain a separate group. Those who could settled in the lowlands, but many were forced to farm the marginal uplands and remote coasts. There, on land scorned by the Punti, the enduring Hakka cut back probably Hong Kong's last remaining natural woodlands.

In 1723 an imperial edict urged that more vacant land in Guangdong be 'opened' and castigated officials for demanding 'squeeze and unofficial fees' from peasants. Offering tax-free earnings from both irrigated and non-irrigated fields, the edict stated:

> *It is shameful to leave fertile land uncultivated. From now on wherever there is waste-land let the people measure it off, open it up and report their claims themselves.*

Hakka migrants were often forced to settle in small, remote valleys such as this one near Double Haven. Encircling Kop Tong village, its fung shui *wood contrasts with the generally denuded hills.*

Four years later, in 1727, an imperial edict was specifically directed at Fujian and Guangdong provinces. Despite good seasons, the edict noted, shortages of rice had occurred due to farmers replacing rice with cash crops: among them, indigo, tobacco, and sugar-cane. The situation—'wealth but no rice'—had been aggravated in Guangdong by some farmers actually deserting their land and turning to trading. The edict, typically North Chinese in its valuing of farming above trading, urged people not to 'neglect agriculture for the sake of making profit. The greatest merit lies in ignoring trivial pursuits and sticking to the hard life of the farmer.'

In reality South China's peasant economies were probably never truly static or 'timeless', and escaping 'the hard life of the farmer' was often seen as highly desirable. Rice remained Hong Kong's staple but, when new crops appeared, there was small-scale experimentation. If a trial crop's growth and profitability seemed promising (and if it needed *less* effort) farmers began planting more of it.*

The Hakka's upland domains reflected this. Hong Kong's fertile lowlands were ideal for 'wet rice' cultivation. But the uplands' poorer, drier land—even where terraces held water and reduced the downhill drift of soil—was better suited to less demanding crops. There, among other crops, the Hakka grew hemp and indigo, maize, peanuts, tea, tobacco, tomatoes, and other vegetables, and rice for village needs. The yields probably fell slowly as more topsoil was eroded, the likely result of the Hakka being forced to over-exploit their scarce land to survive.

On slopes impossible to cultivate, the Hakka still cut the remaining woodland—both for fuel and to gain better access. Mature local woodland was almost impenetrable, and therefore useless. But, once thinned out, half-wild scrubby hills could be used for catching snakes, trapping small animals, and collecting medicinal herbs and fungi. The old village pathways that still cross Hong Kong's hills and mountains show the extent to which isolated villages were linked, and the degree to which the Hakka roamed the hills.

So the seasons, years, and generations passed. The Punti and Hakka clans became ever more deeply connected to the land, their lineages stretching back into an ever more remote past. The Tanka continued fishing, enduring both the sea's dangers and the land dwellers' disdain.

* *Maize, and probably other 'New World' crops, had reached South China during the sixteenth century. The proximity of Hong Kong to Guangzhou gave local farmers early access to the new crops that world trade introduced to China.*

How can one picture the local environment and its people some two centuries ago? The *Xin'an Gazetteer*, whose last edition appeared in 1819, helps frame the picture.

Hong Kong's population then can only be estimated. However, it may have been as large as 50,000. In 1819 there were 854 villages in Xin'an County, at least half of them in Hong Kong—and a census in 1911 gave the population of the mainland New Territories at 69,122, in 598 villages.

Most people were poor, but there was a scattering of prosperous clans and families. Closely related households formed clans and lineages, whose leaders held much authority. Inter-village feuds were common, often due to disputed land-use rights or ownership—and to rival *fung shui* desires. Banditry and piracy were increasingly common, partly due to the weakening authority of the Qing Dynasty.

Belief and superstition permeated the hills and valleys. Besides the all-pervasive implications of *fung shui*, most striking landscape features—rocky outcrops, unusual slopes, certain trees—were invested with meaning. People made offerings there, as they did at countless boulder and banyan shrines dedicated to local earth gods. *Heung-tsuk*, or 'village custom', gave abiding significance to more formal religious observances—for, in fulfilling them, people affirmed their ancestral ties. Festivals were bright interludes in the drab rural round.

Especially in more remote villages, superstition prompted fears of evil spirits—the lurking harbingers of illness, death, or other misfortune. As David Faure eloquently wrote:

> *It cannot be over-emphasized that in the village world the spirits [were] real.... Stories of the sighting of ghosts abound in the village. They appear in dreams, in the brush of wind amongst the trees, in the flickering shadows in the dark.*

Illiteracy was the rural norm, but centuries-old knowledge still structured country life. Village teachers were valued scribes. Bright young men, if they finally became scholars or entered the civil service, brought pride and prestige to their clans and villages.

Esteemed by all, passed on by the literate few, village learning held both proven wisdom and old, unchallenged dogma. The Chinese rural calendar, for example, rested on both tested experience and superstition. And it was the *Farmer's Almanac*, the synthesis of the Chinese lunar, solar, and astrological calendars, that ordered every aspect of the farming and fishing cycles. As Armando da Silva says, 'Expounded in rolling, sonorous verse by a fortune teller or *fung shui* master, [the *Farmer's Almanac*] gave villagers a feeling of immense cosmic significance.'

Virtually every day-to-day necessity was a product of nature. Three styles of hat could be seen around Hong Kong, but all were made from rattan. There was the upswept pointed Punti hat, the flat black-fringed Hakka hat, and the round-domed Tanka hat.

The richest fields and villages were on the mainland's fertile lowlands. The only true market towns were there too, at Tai Po, Yuen Long, and Shenzhen. Among the paddy fields that spread across the valleys were numerous villages, both Punti and Hakka: a few stood alone, some were walled and towered, but most were in clusters. There, and in the nearby fields and hills, most people lived out their lives.

Looking down from the uplands were more widely scattered villages, most of them Hakka. There life was harder, but the homes were like most in Hong Kong. Inside the humble entrances, beneath fir-pole roofs, stood life's necessities: bamboo poles, wicker baskets, and wooden buckets, harrows, ploughs, rakes, and spades. Fan-palm capes and hats hung from lime-washed walls, firewood was stacked near stone grinders. Back rooms had a few frugal comforts: bamboo-framed beds, wooden chests, stools, and shrines.

The coasts and islands were the most remote, barren, and windswept. Their villages were fewer, their temples poorer, their people almost universally illiterate, and their life coarser. Both land and sea were exploited. The women mostly farmed, often on pockets of reclaimed land. The men mostly fished, sometimes using stake-nets but more often small boats.

In 1838 Auguste Borget, a French artist, visited the second largest of these rugged islands (now Hong Kong Island). An anchorage named Heung Gong (now Aberdeen), lay on the island's southern side. It was August when Borget landed there. The summer monsoon was hot and humid. Steep paths led up from the valleys into green-grassed, mostly treeless hills. Borget wandered at will, his eye for telling detail absorbing the land and people. In one village he sketched 'opposite the temple, under an immense tree'—a Chinese Banyan. In another, probably Wong Nai Chung, he wrote:

> *At my feet spread out fields of green rice. On my left, I could perceive many little hamlets embosomed amongst fine large trees ... while in the distance, and terminating the view, appeared the bay and fine mountains of Cow-loon.*

Momentous diplomatic events were about to overtake Hong Kong. But in 1839, the year after Borget's visit, when members of the Liu clan gathered on a hilltop just south of the Shenzhen River, they had a single overriding concern: the weather. The summer had been unseasonably dry. Their paddy fields were parched and their rice was withering on the stalk. Prayers were said for favourable rains, then a rectangular slab of dressed granite was heaved upright. Its inscription states that it was dedicated to the god of 'rain and cloud'. The stone, raised during the reign of the Qing Emperor Dao Guang, still stands overlooking Sheung Shui.

A grassy hilltop in the northern New Territories.

Dawn—and a cargo ship passes through the East Lamma Channel, delineated by Lamma and Hong Kong islands.

REGION

SOUTHERN ISLANDS

Dawn just breaking and our ship passing the first islands—dark islands in a silvery sea, the colour quickening. Islands, and islands, yet more islands, many of them high and steep and green to the water's edge. Behind them mountains, peak upon peak, range behind range, with one enormous pillar of cloud rising above them all.

MARY C. KNIGHT, 1949

Summer. Dawn light is spreading across the sea, and a line of ships stretches towards the horizon. From my camp high on Lamma Island the vessels seem toy-like. In fact they are massive container ships, totally different from the passenger ships of the 1940s.

A curious thought strikes me: these leviathans, steaming into Hong Kong with the first light, carry virtually no travellers—just hard-working crews. When the missionary Mary Knight arrived here in 1949 it was different. As her SS *Canton* steered towards the harbour, like countless other ships before and after, Mary's fellow passengers doubtless watched spellbound as Hong Kong revealed itself.

Their approach lay among some rugged islands which, like giant stepping stones, encircle Hong Kong Island to the east and south. Former Ice Age peaks sloping up from a coastal plain, after the sea level rose they became islands—and so shaped the channels leading towards Hong Kong. For many centuries sea passengers passed amid these rocky landfalls; now air travellers barely glimpse them as they hurtle down towards yet another airport.

Across waters sometimes blue-green, sometimes gun-metal grey and sometimes golden, this small archipelago stretches for some 30 kilometres. Treacherous crags, numerous tiny islets, and some larger islands make up its parts. Of the latter, the Ninepin Group and Tung Lung Chau lie to the east of Hong Kong Island; Waglan Island, Po Toi, and Beaufort Island stand to the south-east; and Lamma Island lies to the south-west. Further out are true ocean islands, but these never formed part of the Colony of Hong Kong.

Castle Rock, typical of the tide-washed crags, rises just a few metres above the sea. Waglan, about 50 metres high, is little more than a jagged strip of rock. But Tung Lung and Po Toi, each about two kilometres across and over 200 metres in height, give commanding views. Lamma is by far the largest of these islands, and there Mount Stenhouse rises to over 350 metres.*

Surprisingly remote even today, these islands are most commonly seen half-hidden beyond headlands from Hong Kong Island. The sea has largely insulated them and, excluding parts of Lamma, they are virtually as they have been for centuries. Tung Lung Chau reveals the ocean's awesome power. There and on Po Toi, rock engravings reach back into misty antiquity—and on Lamma archaeology has brought Hong Kong's prehistory vividly to life. Significant village communities have grown up on Lamma.

The islands and islets of the Ninepin Group (Kwo Chau Islands) are, as their name suggests, steep and sharp—and with impressive cliffs. Indeed, so rugged are the Ninepins that it is almost impossible to land on their uninhabited islands. Cliff-devotees must visit Tung Lung Chau.

Seen from above, Tung Lung Chau (East Dragon Island) looks more like a turtle than a dragon. Flippering southwards into Tathong Channel, its head is the bulge of Tathong Point, its tail the rocky promontory facing back towards Joss House Bay. The island's western side suits the image, as there it descends gently to a rounded coastline where its few people live. But Tung Lung's eastern face destroys the tranquil picture: for there it plunges violently into the ocean.

* *Hong Kong's waters are much shallower than the plummeting slopes of these islands suggest. The deepest channels around Lamma and Po Toi are little over ten fathoms, or about 20 metres.*

The vast majority of Hong Kong's coasts are steep and rocky. Even so, Tung Lung stands apart. Nowhere else in the Territory are there cliffs so high or so forbidding. At one point the land plummets 200 metres in the same horizontal distance—a drop equivalent to about half the height of Central Plaza, Hong Kong's tallest building, across the width of a few city blocks.

Relentless swells have carved out these islands' cliffs. In heavy weather, as unimpeded ocean swells meet these eastern islands, air trapped between solid rock and surging water is massively compressed. The split-second of impact can produce pressures upwards of 150 kilograms per square centimetre, roughly similar to that of two medium-sized adults standing on a stamp-sized area. Imagine a rock face some ten metres square: a similarly powerful swell would crash over it with an overall force of up to 150,000 tonnes!

Accept that—listen for the uumpphhhh of an attacking swell—watch the fearsome suction as the ocean draws back—and one has some sense of the ocean's erosive power. Rock, hard as it is, slowly, inevitably disintegrates before such forces. Even on 'calm' days ocean swells roll in past the Ninepins, often strong enough to cause smaller ships to pitch and to turn junk picnickers green.

Below these cliffs marine creatures maintain a precarious foothold, hidden in crevices, secured to the least exposed rocks. Adapted to being constantly washed over, specialized crabs, barnacles, mussels, and limpets somehow avoid the worst explosions of storm-driven waves. Along Tung Lung's sheltered eastern side, as on the other islands, occasional bays and coves have developed. Here the coastal scene is more varied. Boulders, often resting on sand, offer promising niches for many more crustaceans and fish. Beneath the sand animals burrow and forage: clams, cockles, fan shells, worms, shrimps, and snails.

The safety that the islands' sheltered bays offered—and their relatively easily procured food—no doubt encouraged the seafarers who first reached Hong Kong some time after 5000 BC.

On Tung Lung Chau and Po Toi Island long-gone people carved

On the east coast of Tung Lung Chau, ocean swells have carved some of Hong Kong's most forbidding sea cliffs.

Mat Chau, seen from the bay where Po Toi's rock carvings are found.

designs into rocks facing the sea. The Tung Lung carving looks towards Lei Yue Mun (Carp Gate or Strait), the narrow eastern entrance to Victoria Harbour. Near a small sea cavern some geometric designs are cut into a rock slab. Some historians believe they date back only to the Song Dynasty (960–1279), but others believe they are prehistoric. The rock carvings are marked on the *Countryside Series* maps.

The carvings on Po Toi, 10 kilometres south of Tung Lung, are almost certainly prehistoric. Near Po Toi's single small village, at its south-western angle, is Nam Tam Wan. There, beneath slopes heavy with granite boulders, on a rock face that drops into the water, designs similar to Tung Lung's can be seen. Across the bay is a craggy promontory scattered with ancestral graves, for propitious *fung shui* their backs to the hillside and their faces to the bay.

The rock carvings' precise age and purpose can only be guessed at. But they may well have been carved by Yueh tribes, the people who probably first settled among Hong Kong's islands some six thousand years ago. One thing is certain: the carvings must have needed time-consuming effort, and so were probably prompted by strong spiritual or magical beliefs; perhaps they were made to placate the ever-dangerous sea.

Across a narrow channel from Po Toi, Beaufort Island (Lo Chau) rises abruptly. Hardly surprisingly, Beaufort has no carvings for the island is incredibly precipitous. Massive rock sheets slope straight into the sea without a fragment of foreshore. Landing is virtually impossible and there is no evidence of anyone ever having lived there.

The Tung Lung and Po Toi carvings only hint at antiquity. The remains unearthed on Lamma Island, however, present a rich impression of Hong Kong's distant past. Lamma (South Fork) is also the loveliest of these islands: sinuous and deeply indented—in places steep and rugged, elsewhere low and homely—and always full of charm. Not surprisingly, it is the only one of these islands with a significant, but small, population.

Grand but not inhospitable, Lamma is by far the largest of the southern islands. This view, from above Yung Shue Wan, looks south towards Mount Stenhouse.

I admit some bias. I liked Lamma when I first visited the island, and moved there in 1993. Lamma's rugged beauty helped spur me to write this book: its country was striking enough—and perhaps threatened enough—to be worth recording.

Only the Black-eared Kites that wheel above Mount Stenhouse see the full complexity of Lamma's orchid-like outline—no other Hong Kong island is so intricately shaped, except the much smaller Crooked Island. Lamma covers almost 14 square kilometres, and at its longest spans seven kilometres. Peninsulas extend from its centre, enclosing deep bays facing this way and that. The valleys are small, but fertile and well-watered. Trees are few, except on the lower slopes and valleys, due largely to Lamma's exposed situation—as is the case on all the islands.

Lamma is also ideally placed: midway between the ocean to the east and the estuary to the west, and close to Hong Kong Island. With Lamma beckoning, why settle on Tung Lung or Po Toi?

It is thus hardly surprising that Lamma was settled early, as a glance at a local prehistorical site map confirms. By far the largest number of Hong Kong archaeological discoveries have been made among the southern islands, and Lamma boasts the greatest number of excavations.

By 1930 archaeologists had already established that human settlement around Hong Kong extended far back. However most prehistoric remains had simply been picked up on eroding hillsides, and their dispersal by typhoons and heavy rain meant that dating them accurately was impossible. Various pottery styles and stone artefacts, however, indicated distinct periods: the earliest remains pre-dated the Chou Dynasty (722–222 BC).

Enter Daniel J. Finn, a Jesuit and lecturer in Geography at the University of Hong Kong. C. M. Heanley, J. L. Shellshear, and W. Schofield, pioneer Hong Kong archaeologists known as 'The Honorary Society of Beachcombers', had been urging Finn to join their excursions. He was undecided. Then one day in the early 1930s, where sand was being unloaded at an Aberdeen jetty, Finn almost stood on some prehistoric pottery shards. As he later wrote:

> *On my way back past the same spot, I picked up a small piece of pattinated bronze, evidently a fragment of a weapon. Next morning, I came deliberately with a friend and a shovel. My friend had the satisfaction of digging up a stone spear-head weathered to a russet-brown, and we got more fragments of pottery.*

Finn's curiosity was aroused. Where had the material come from? Lamma, the labourers replied—the island that lay just three kilometres south of Aberdeen, across the channel.

Finn began scouring Lamma's foreshores. Modest and amiable, then in his forties, Finn dug enthusiastically into the puzzle of Hong Kong's past—and into Lamma itself. In a few short years during the 1930s he made invaluable discoveries, and wrote numerous articles on Hong Kong archaeology. Finn's book, *Archaeological Finds on Lamma Island*, was published posthumously in 1958, edited by Thomas F. Ryan, another Jesuit.

Tai Wan (Big Bay), a fertile Lamma valley, was Finn's first site. Its abundant water and fields spreading up into terraced hills suggested a likely settlement area. Pith helmeted, his tropical whites held up with braces, Finn stood by while labourers sieved sand to remove coarser gravel—and, in the process, occasional prehistoric shards. 'There is no concern for what may be mixed up in the sand unless it gets in the way', Finn regretted. 'Small objects such as the bits I picked up at Aberdeen can only by the merest luck escape removal and dispersal.'

In 1933 the government financed Finn's proper excavations at Tai Wan. He later also excavated at Yung Shue Wan (Banyan Tree Bay) and Hung Shing Yeh, other valleys at Lamma's northern end. Finn had a good rapport with the Lamma people. But the villagers, concerned lest his excavations might disturb their fields' *fung shui* (and trouble departed spirits), later discouraged the digging.

The sheer variety—and fragmentary nature—of the stone, pottery, and bronze remains unearthed was daunting. But Finn brought his wide-ranging Asian scholarship to bear. And, though he admitted it was 'mere conjecture', he concluded that the earliest Lamma remains might have been left by sea-faring tribes who came from around Indo-China. However, aware how much remained to be done, Finn admitted:

> *It is an obvious fact that the archaeology of these regions is yet far from being clear, and it is patent that the Lamma [and other] Hong Kong finds add their quota of mystification.*

Finn died suddenly in London in November 1936, but his archaeological work endured. Thirty-five years later, in April 1971, another Lamma site yielded the first of its remarkable prehistoric remains. The objects spanned six millenia—back to about 4000 BC—and largely confirmed Finn's 'mere conjecture'. These unprecedented finds were unearthed at Sham Wan, a bay on Lamma's south-eastern side under Mount Stenhouse.

It is autumn, and I am scrambling up from Sham Wan's bay towards Mount Stenhouse. Reconnoitring this track two months before had taken less than an hour, but now ninety minutes have passed. Despite the forecast—'a fine afternoon'—rain has left me soaked. The track, overgrown with summer grasses and cut deeply into the clay, is running with water. Exhausted, I finally reach my objective: a flat-topped granite outcrop which gives commanding views.

It is almost sunset. Sham Wan's deep-set bay lies far below, its crescent beach backed by the rising ground where prehistoric discoveries have been made. Looking along Mount Stenhouse's side, my gaze moves over an untrodden wild of grassy boulder-strewn slopes. Further out the coast, edged with plunging rock, is completely inaccessible.

How well Sham Wan suited its early people, I reflect. An enclosed bay, protected by long high arms. A peak nearby, with gullies and springs for water. A beach for easy landing and boat repairs, and with driftwood for fires. Behind the beach, some higher flat land. And around the bay boulders to fish from, and to collect shellfish beneath.

Later a full moon rises above Po Toi Island, 15 kilometres to the east. Silver lacework spreads across the sea. Down near Sham Wan pinpoints of light appear and then disappear—shifting moonlight reflecting on the water in some rockholes.

Midnight. Waves wash on to Sham Wan; the air is close. A sudden rain squall forces me under a tarpaulin, but driving rain seeps in. The remainder of the night is punctuated with downpours and, now wet through, I hardly sleep. Finally, an hour before dawn, a thunderstorm advances ominously. I pack my rucksack, ready to abandon the lookout if lightning forks down too close.

At last daybreak comes, its light soft and diffuse. Heavy clouds roll in from the South China Sea. They approach in solid banks, hide Mount Stenhouse in swirling vapour, descend lower, envelop me in desolate grey. The raw elements: the same forces that long ago often must have assailed Sham Wan's first people.

❦

The settled villages on Lamma, Po Toi, and Tung Lung have been there for centuries. Largely self-sufficient and rarely venturing to the mainland, the islands' Punti and Tanka people tilled the land and fished around the islands. The populations were small, the resources taken from the land and sea were largely sustainable—and so the islanders caused only minor environmental change.

Exploring Lamma in the early 1960s, the poet Edmund Blunden found fishing-farming communities still remote from the changes then overtaking urban Hong Kong—indeed, virtually cut off from the twentieth century. Around Lamma's scattered villages and fields, life remained patterned by the Chinese calendar, the seasons, and the tides. Lamma's wilder parts were mainly untouched, as they still are.

Further prehistoric sites had been discovered on Lamma by the 1960s, gradually clarifying the prehistoric story. Moved by witnessing an excavation, Blunden, wrote:

Beneath the old calm sky. We cannot hold
In hands untrembling these skilled works of old.
He takes them gently, richly, like red gold.

Archaeological discoveries on Lamma now have been made at some ten sites, of which Sham Wan is pre-eminent. Pre-war surface finds there had indicated that the bay's raised sand bank was an early living site, but it was the archaeologist William Meacham who first proved Sham Wan's potential. Meacham and Solomon Bard, helped by Lamma villagers, subsequently led Sham Wan excavations between 1971 and 1977.

Until then, at no other Hong Kong site had clearly dated finds of such antiquity been made. And nowhere else had excavations revealed such continuous settlement. The Sham Wan excavations pushed back Hong Kong's known human occupation by some 1,500 years, to about 4000 BC. The site brought to life each main phase of Hong Kong's past, spanning some 6,000 years: the historic period, reaching back to about 200 BC; the Bronze Age, to about 1200 BC; and the Neolithic period, to about 4000 BC.

Trowel load by painstaking trowel load, the Sham Wan artefacts of one period gave way to those of the previous one. Burial remains were found for each period: a Qing coffin and teapots; two Tang burial jars; a Bronze Age stone slab; and various Neolithic funerary objects. There were human remains from the earliest periods, some indicating cremation; and food remains, especially fish bones and oyster shells.

❦

Island people, mesmerized by the sea horizon yet forever gazing inward at themselves, tend to shun their wider region. Hong Kong's islanders were no different.

Some, encouraged by their indented archipelago, turned to piracy. For centuries attacks flared with an almost seasonal regularity, as pirate junks sallied forth with the winter monsoon. Chinese governments intermittently tried to bludgeon or at least control them; but the pirates, hidden in remote bays and coves, often simply waited. The winds would always return: their

Po Toi is a true ocean island: rugged, wild, and mostly inaccessible. These rock-faces line its south-eastern coast.

Wearing away the exposed side of Mat Chau, an islet just off Po Toi, waves have created these low cliffs.

Most of the southern islands are virtually uninhabited. Only Lamma has large communities, at Sok Kwu Wan and Yung Shue Wan—where this power station stands.

pursuers might not. Well into this century pirates were still active in Hong Kong waters, and smugglers remain a problem today.

Other islanders turned to peaceful lives at sea. When Britain colonized Hong Kong Island, the Lamma villagers let it be known that—with the Aberdeen fisher-folk—they wished their island to be part of the Colony. This did not occur until 1898, but Lamma men went to sea on British merchant ships. Even today English nautical terms are used by some older Lamma Island men.

Was it a wish to be left alone, safe from both officials and pirates, that led some people to settle far from the main island villages? Or was it the sheer shortage of usable land that led them to build hamlets where outside contacts must have been minimal? Whatever the reason, scattered through the islands in remote places, one sees decaying cottages, the remains of terraces, and long-forgotten *kam taap* (burial urns). Some such places have not been long abandoned, and yet even so the vegetation has grown back and smothered the remains of old buildings.

Indeed, given half a chance, nature grows back with a vengeance in Hong Kong. But nothing can grow when there is no soil or even rock. And, tonne by tonne, even large islands can be literally blasted into the sea. If Chek Lap Kok is not evidence enough of this, visit the Chinese Zhizhou Islands south-east of Lantau (or look at them on the way to Macau by hydrofoil). The sheer quarrying-away of these high, rugged islands is staggering—and of great environmental concern. Much of the resulting landfill is destined for Hong Kong.

Until now the Hong Kong islands described here, excluding Lamma, have been mostly left untouched. On Lamma, there is now a power station and a major quarry. The Yung Shue Wan power station was very well-sited and barely intrudes, but the Sok Kwu Wan quarry is a visual eyesore. Meanwhile, 'villa' flats are spreading rapidly around Yung Shue Wan with little apparent planning—and, like a proposed reclamation there, with no overall vision to complement the natural landscape.

Despite Lamma's size and beauty, the island has not a single hectare of Country Park. Little-known plans exist to create one, and proposals have been put forward by the Country Parks Board to the government for some years. Establishing a Lamma Country

An isolated Lamma hamlet, battered by wind and spray during a tropical storm.

Acorn Barnacles at Sham Wan on Lamma Island. Prehistoric remains discovered nearby show that shellfish formed part of the diet of Lamma's first people.

Park would require only about five million dollars, with annual running costs of about half that. Yet each year the government has replied that the necessary funds cannot be found. Given Hong Kong's wealth, and its financial reserves, this can hardly be believed. Indeed, whatever the complexities of budgetary pigeon-holes, it can only be seen as a lame excuse.

Until such a Country Park is declared, it cannot be assumed that Lamma is safe from development. Most of the island is too rugged to build on, but the shallowness of its bays means that major reclamations are always possible.

Among these rocky islands, at night and especially when typhoons threatened, fishing craft and seagoing vessels alike were always in danger of running aground—as, to a lesser extent, they still are.

Ask any China coast navigator (ideally an old salt whose life was spent with lead lines and sextants) to name a Hong Kong landfall, and the likely reply will be Waglan Island.

Waglan, in fact two adjacent islets, is mostly so low that during storms its rocks are almost hidden under flying spray. But barren and windswept as Waglan Island is, in 1893 the Imperial Chinese Maritime Customs built a lighthouse there. The island came under British control with the ceding of the New Territories in 1898, and its light joined countless others across Britain's sea-knit empire on 1 January 1901. The Cape D'Aguilar light, on Hong Kong Island opposite Waglan, was then closed.

For a century Waglan's finger of steel and concrete has survived the lash of typhoons. And every night, without fail, Waglan has marked the position 22° 11.10′ North, 114° 18.1′ East by flashing 'gp.fl.2 w/30 sec'—two white flashes sweeping across the sky every thirty seconds, seen in clear conditions 16 kilometres away.

The Waglan light, though no longer manned, is still by far the brightest beacon around Hong Kong's waters. Most ships still approach Hong Kong from the ocean to the east, and Waglan remains their first landfall. Daily, on the bridges of the many ships threading a passage through the outlying islands, Waglan is framed in binoculars: barren, windswept—and comforting.

In the 1840s Hong Kong Island's treeless slopes fell to a narrow coastal strip. Excluding its top profile, the Island's northern side looked much like this— the Sharp Peak peninsula today.

HISTORY 1841–1900

THAT BARREN ROCK

'A pox on this stinking island!' Brock said, staring around the beach and up at the mountains. 'The whole of China at our feets and all we takes be this barren, sodding rock.'

JAMES CLAVELL, *TAIPAN*

The more I saw of Hong Kong, the more I was struck with the enterprise which had created such a city on such a spot .... Roads had been cut out of the solid granite, and the trees visible over the face of the mountain had all been planted in equally unpromising soil.

SIR WILLIAM DES VOEUX, 1903

In the young colony of Hong Kong the summers were a misery. There had, of course, been no change in the seasons. But Europeans, unused to subtropical heat, now lived there. And for them the southerly monsoon brought not just humid heat but scorching days and oppressive nights. True, there were delightful winters to anticipate. But during the long summers it was hard to believe that the cool, dry months would ever come.

Even in the late 1880s, when living conditions were far better than forty years before, most Europeans found the summers unbearable. For a privileged few, the Peak on Hong Kong Island held some relief: there the air was a few degrees cooler, but the place was perpetually moist.

Governor Sir William Des Voeux, for one, was uncertain whether the Peak or the harbour city of Victoria was less trying. During one summer the walls of Mountain Lodge, the Governor's Peak residence, actually dripped with water. Des Voeux felt trapped there in his 'damp and gloomy prison'. He and his wife occasionally escaped down to Government House, near Victoria. But there they suffered another climatic torment: 'It was a pleasure to see daylight and bright sky again, but the heat, which rendered sleep almost impossible, quickly drove us up again.'

'If you live on the Peak your clothes rot; it you live below, you rot', one resident concluded dismally. Most people, however, had no choice. Europeans and Chinese alike, they simply stayed down around the harbour, sweltering through the hot wet summers.

The land that today comprises Hong Kong passed to Britain in three parcels. Hong Kong Island, and later Stonecutters Island and most of Kowloon, were ceded to Britain in 1843 and 1860 respectively. The New Territories, including the other islands, were leased to Britain in 1898 for 99 years.

The six decades from 1841 to 1900 saw Hong Kong dramatically changed. In 1841 the landscape was half-wild, half-agricultural; by the end of the century the city of Victoria faced the harbour. The Colony's growth as a trading port was centred on Hong Kong Island. In 1841 Hong Kong Island's village population was about 4,000, with perhaps another 2,000 living on boats. By 1845 it had already passed 23,000, ten years later it was over 72,000, and by 1886 it had reached 180,000.

On the mainland and the islands the old farming-fishing life continued largely unchanged. But on Hong Kong Island, and later Kowloon, the fast-growing urban population led to ever-greater incursions into the countryside. Most people were too poor (and some too obsessed with becoming still wealthier) for the surrounding natural grandeur to touch them. But even the most driven souls probably marvelled at the development spreading ever-further around the harbour and up the slopes. Further afield on Hong Kong Island, reservoirs later took the place of valleys and reforestation crept up the hillsides.

Europeans, aided by trading profits, Western technology, and British administration, remoulded the Hong Kong environment. But the Colony's climate and terrain, as extreme as before, were demanding. Most Europeans found Hong Kong exhausting, and for many it brought sickness and death.

The overwhelming majority of Hong Kong people were Chinese: the local Punti, Hakka, and Tanka, and countless

Insects and spiders added to the trials of tropical life. Woodland Spiders, such as this large female, often hang in wait along pathways—corridors for insects as well as humans!

newcomers from China. Some Chinese prospered in the British haven. But most, though better off than in China, endured lives of poverty and toil. Yet it was these poor sojourners who, with bamboo poles and wicker baskets, physically transformed Hong Kong—carrying the Colony across an historic watershed, from its agricultural past towards its urban future.

Well before 1841 British ships had often called at Hong Kong Island to replenish their water. Clark Abel, Surgeon to Lord Amherst's ill-fated Embassy to the Emperor of China, observed in 1816 that Hong Kong Island was 'chiefly remarkable for its high conical mountains, rising in the centre, and for a beautiful cascade, which rolled over a fine blue rock into the sea'. The scenery was mostly 'barren rocks, deep ravines, and mountain torrents', the only inhabitants were 'poor and weather-beaten fishermen'.*

* *A common water supply point was the waterfall near the fishing village of Heung Gong (Fragrant Harbour), named after the wood of the Incense Tree exported from there.*

Britain claimed Hong Kong Island in 1841, but the treaty granting it was not ratified until 1843. Dwarfed by towering peaks, a British naval party raised the flag at Possession Point on 26 January 1841. Above them Hong Kong Island's sheer western slopes rose up, brown with winter grasses and dark with rock.

When Captain Edward Belcher charted Hong Kong Island's coastline in January and February 1841, the Island's peaks seemed its most prominent features—and Belcher named almost all of them. Hemmed in by the harbour and peaks, a Madame Ida Pfeiffer found the foreshore 'not very pleasing ... [and] surrounded by naked hills'. The Revd G. N. Wright, whose text accompanied T. Allom's landscape engravings published in 1843, noted that 'although beautiful in the distance ... [Hong Kong Island] is sterile and uncompromising upon a more close examination'.

Even allowing for some romantic exaggeration, most colonial artists were clearly struck by Hong Kong's precipitous, rocky aspect. To them, the Island was dominated by jagged peaks and deep ravines. The sheer slopes eased to moderate inclinations only near the sea, and the harbour was indented with promontories, coves, and inlets.

Ironically, the very feature that was to ensure the Colony's future prosperity—its harbour—faced the Island's rugged northern side. As Lieutenant Thomas Collinson R. E. found when he meticulously mapped Hong Kong Island between 1843 and 1845, the Island's southern side was far more varied and beautiful. The southern villages were also larger, and 'exceedingly neat in appearance with blue-tiled and white-walled houses'.

Everywhere the hills were mostly treeless, and the only 'fertility' was in some small agricultural valleys. Above Hong Kong Island's 'rice grounds', Allom found many 'romantic little glens ... adorned with masses of rock, in the fissures of which the noblest forest trees have found sufficient soil'. And the Revd Wright wrote, 'Advantage has been taken by encouraging the growth of timber in the glens, within which the loveliest hamlets may be seen enclosed'. He added: 'Industry, unequalled in any other kingdom, has converted a soil the most discouraging into one the most productive.'

The 'barren' hills were, of course, the result of long-term deforestation by man, as Robert Fortune, the British botanist, probably realized. Fortune passed through Hong Kong in 1844, before leaving to begin his great botanical work in China. The Hong Kong hills, he observed,

> *have everywhere a scorched appearance, with rocks of granite and red clay showing all over their surface. The trees are few, and stunted in their growth, being perfectly useless except for firewood, the purpose to which they are generally applied.*

Despite its deforested hills, botanists discovered that Hong Kong has a diverse flora. Ixora chinensis, *seen here, was noted by Robert Fortune in 1844.*

The soil was thin, especially on the eroded hills. Fortune knew that improving the vegetation would thus be difficult, but he noted that some Europeans had already begun gardens. Trees—and shade—were sorely needed. Without them, as Fortune wrote, the Hong Kong sun had 'a fierceness and an aggressiveness ... which one never experiences in any other part of the tropics'. And he urged the government to 'use every means in their power to clothe the hillsides ... with a healthy vegetation'.

There is no escaping the impact of Hong Kong's hot, humid summers on Europeans. George Wingrove Cook, *The Times's* rather acerbic correspondent, noted that Victoria was shut off by the hills from the summer southerly winds. Thus, the heat there was 'a stagnant, up and down, fierce, often reflected heat—a heat there [was] no escaping'. People often woke to hear cascading rain and, once the sun rose, Victoria became 'one hot vapour bath'. Even at one's club, dressed for dinner, the summer monsoon brought stormy interruptions. As Cook recorded:

> *The doors slam and the verandah blinds clash, rheumatisms and agues riot boisterously about; while in mockery of the windy turmoil the coolie, crouched in one corner of the room absorbed in the ecstasies of an opium dream, continues his gentle pull at the now madly swaying punkah.*

Ants and cockroaches added to the trials of domestic life. And there were mosquitoes, which as Cook wrote, 'insinuated themselves through your fortress of gauze'. Along the pathways that wound through Victoria's still half-wild hills there were spiders and snakes.

Between 1841 and 1845 Hong Kong experienced its full range of climatic extremes: typhoons, devastating rain, and water shortages. A severe typhoon struck in July 1841. Captain Charles Elliot was shipwrecked on some nearby Chinese islands; and around Victoria Harbour the temporary bamboo-framed buildings were flattened. Four years later, in May 1845, rainstorms created torrents that swept some Chinese labourers out to sea and swamped Victoria with soil and debris. The *Hong Kong Register* reported:

> *Houses were undermined; roads, made at great expense only a few months before, were swept away; drains were burst open; and many of the bridges and other public works rendered perfectly useless.*

Hong Kong's granites, such as this scalloped outcrop, yielded valuable stone for buildings.

Typhoons and rainstorms were occasional ordeals. But the summer heat and humidity, and the disease they appeared to bring, were constant reminders of the extreme environment. Mortality, especially among Europeans, was appalling.

'Sickness and death lurk amidst picturesque scenery', wrote Henry Charles Sirr, a lawyer, lamenting the death toll every summer in the late 1840s. Damning Hong Kong Island as 'pestilential' Sirr continued: 'the exhalations arising from the water produce fever and ague which too frequently terminate fatally'.

The sickness of Europeans was in part due to their being unused to the heat—and to heavy clothing and poor hygiene. But another reason was the actual development of Hong Kong. During the summers, quarries and building sites provided countless rainfilled areas for anopheles mosquitoes to breed in—and, unbeknown, the mosquitoes spread malaria. The already over-crowded urban areas also hastened the spread of disease, through festering refuse, flies, and rats.

The grim saga of sickness and death that marred the Colony's first decade cannot be pursued here, but two facts hint at its tragic scale. In 1848, 20 per cent of the British troops in Hong Kong died, compared to only about 1 per cent of the Chinese population; and in the nine years from June 1849, the 59th Regiment lost some 700 men, women, and children.*

Hong Kong Island, where it faced the harbour, was composed almost entirely of granite. Granite outcrops framed the hills, and beneath the gravelly slopes lay countless tonnes of weathered granite. Among the first Chinese workers to reach the Colony were Hakka stone-cutters, sinewy labourers whose tools of trade were pickaxes, sledgehammers, and stone chisels. As the historian G. R. Sayer wrote: 'At the magic touch of British capital and

* *Harrowing accounts of the 59th's suffering, recorded by Private James Bodell, have recently been published in* A Soldier's View of Empire, *edited by Kevin Sinclair. Bodell wrote of the summer of 1850: 'During July, Augt and September, we buried about 300 men. I never seen or heard anything like the Epidemic that got amongst the men....'*

Chinese labour, Hong Kong's unprofitable hills yielded up their hidden treasure and a town of native granite emerged'. Not for nothing was the word 'coolie', now regarded as derogatory, derived from the Chinese *ku li*—meaning 'bitter strength'.

As early as the mid-1840s the harbour's smaller promontories had been quarried away, and its beaches and coves were disappearing beneath stone walls and quays. The lower hillsides were being terraced, to make level foundations for buildings and to yield granite for the city of Victoria. Among solid granite buildings pathways wound towards open hillsides.

In 1845 Lieutenant Thomas Collinson drew a panorama (now in the Museum of Art) looking westwards over Causeway Bay. His topographical drawing, a cartographer's unembellished work, no doubt provides an accurate impression of how Hong Kong Island then appeared.

The higher hills and peaks were untouched, their ridges and gullies barely accessible. Set between low spurs, Causeway Bay and Happy Valley were still covered with paddy fields. Low-lying Causeway Bay was separated from the harbour by a bund or dyke (hence its name). But many of the hillocks near the harbour had had their summits razed, and further west buildings lined Victoria's foreshore. Across the harbour the Kowloon Peninsula was a maze of hills; Hung Hom was an expanse of paddy fields; and a beach curved round from Tsim Sha Tsui to Yau Ma Tei. At Happy Valley Collinson had pencilled in an ominous note: 'The New Grave Yard'—the Colonial Cemetery laid out in 1845.

During the second half of the nineteenth century turmoil in China rapidly increased the flow of émigrés to Hong Kong—so affecting its environment. The Tai Ping Rebellion (1850–64) brought chaos to China. At least twenty million people perished; and countless refugees fled to Hong Kong. Between 1853 and 1865, the population tripled from about 37,000 to over 125,000, with only some 2,000 Europeans. By 1865 over 4,500 ships were berthing in Hong Kong each year.

Stonecutters Island and the Kowloon Peninsula became British territory in 1860, by the Convention of Peking. However, the overwhelming majority of people still lived on Hong Kong Island. Victoria had an increasingly European appearance, but its British community—often divided and vexatious—was largely a veneer. Hong Kong's enduring reality lay in its teeming Chinese streets, mostly inhabited by poverty-stricken émigrés.

The Colony generally attracted South China's least fortunate people: labourers, hawkers, servants, prostitutes, criminals—as Governor Sir Richard Macdonnell (1866–72) lamented, many were 'the moral refuse of Canton'. Inured to bitter hardship and corrupt officials in China, these new arrivals resisted any Hong Kong government involvement in their lives. They grasped the opportunity for bettering their lives that British order gave them, yet mostly resented the fact of British rule. For their part, colonial officials often seemed little troubled by the Chinese tenements' conditions: squalid over-crowding, unsewered latrines, polluted wells, decaying filth—and the escape into opium addiction.

By the 1860s, as paintings show, barren Hong Kong Island had been greatly changed. Its higher peaks were still empty, but beneath the lower slopes Victoria had a cluster of impressive buildings. There was no longer any vacant land and, defying the steep terrain, residences had reached the 'mid-levels' (buses careering downhill through Mid-levels today emphasize the difficulties of building there). Further west tiers of Chinese tenements spread up from the waterfront. Unseen but invaluable, underground 'nullahs' and storm-water drains now led from the hills to the harbour, freeing the city from torrents of silt after all but the worst rains.

Happy Valley's paddy fields and swamps had been drained and reclaimed, providing Hong Kong Island with its only extensive flat land—grassed over for a race course. Elsewhere there were always hills, the preserve of stoic sedan-chair bearers 'never weary of climbing the steep and tortuous streets, or of [winding up] the scorching pathways'. The only way into the mountains was along village pathways, and only intrepid individuals climbed them. Laurence Oliphant did not. Sweltering aboard a P & O ship in port during September 1862, he yearned for sufficient energy to get ashore—let alone beyond Victoria:

> *Hong Kong boasts only two walks for the conscientious valetudinarian, one along the seashore to the right, and the other to the left of the settlement. Then there is a scramble to the top of Victoria Peak, but this achievement involves an early start and a probable attack of fever.*

Despite the urbanization on Hong Kong Island, Kowloon remained largely agricultural until almost the end of the century. The peninsula, as a British Parliamentary Paper noted in 1861, was 'extremely rugged [with] numerous small hills divided by ravines and patches of marshes and rice fields'. The peninsula proper had ten villages and a few thousand people, mostly Hakka vegetable farmers now supplying Victoria, and some Tanka boat people living around Yau Ma Tei. A tall bamboo fence snaked along the border with China (today's Boundary Street), beyond which were some Punti villages in fertile valleys.

The summer monsoons brought enervating humidity, drenching rain, and majestic clouds—such as this monster towering over Mirs Bay.

The summer monsoons also brought tropical storms—as seen here at Cape D'Aguilar.

During the nineteenth century, typhoons sometimes caused tragic loss of life, especially among the boat people.

Besides being regarded as a European preserve, one reason for Kowloon's relatively slow growth was its reputation for sickness. The government's principal medical officer reported in December 1861 that the peninsula was mostly 'intersected by belts of paddy cultivation requiring constant irrigation'. The British regiment that first occupied Kowloon was decimated by 'Hong Kong fever', blamed on gases emerging from the paddy fields—but in reality caused by mosquitoes breeding in the rice fields.

Hong Kong was fortunate in its early European botanists. Those who first explored the Island's hills and valleys found some old *fung shui* woods near three villages: above Heung Gong (Aberdeen), at Wong Nai Chung (above Happy Valley) and at Tai Tam Tuk. Chinese Pines were the only trees on the hills, and fuel-gathering kept them few and stunted. Tree- and shrub-cutting also largely explained why, as Robert Fortune noted, 'almost all the ornamental flowering plants are found high up in the mountains'. Only in the wetter, inaccessible gullies were flowering shrubs, such as *Ixora chinensis*, common. As Fortune wrote in *Three Years Wandering in China*:

> *Higher up in the mountains we find the beautiful* Ixora *flowering in profusion in the clefts.... Its scarlet heads of bloom under the Hong Kong sun are of the most dazzling brightness.*

During the 1840s and 1850s hitherto unknown species were described by the other botanists: Richard Brinsley Hinds, Dr Henry Fletcher Hance, Lieutenant Colonel J. Eyre, Captain John George Champion (and his wife), Governor Sir John Bowring (and his son), and Dr Berthold Seamann. Their discoveries are commemorated in the plants they named, or which were named after them; and their work endures in the pages of *Flora Hongkongensis*—the first full record of Hong Kong's flora.

Published in 1861, *Flora Hongkongensis* was compiled by George Bentham, a brilliant botanist undaunted by never having visited the Colony. *Flora Hongkongensis* runs to over 500 pages, and lists 1,056 species, distributed into 591 genera and 125 orders. As Bentham's Preface observes:

Government reforestation was intended both to 'green' Hong Kong Island and to prevent the reservoirs from silting up. Chinese Pines were the most widely planted species.

> *At a first glance one is struck with the very large total amount of species crowded upon so small an island, which all navigators depict as apparently so bleak and bare ... and with the very great diversity in the species themselves.*

Hong Kong's tropical-temperate climate largely accounts for the great botanical diversity in so small an area.

Hong Kong's Botanical Gardens drew on the experience of local botanists. Begun in 1861, the Botanical Gardens were a major undertaking. The site above Government House had to be terraced out of solid granite—and there was almost no soil. Chinese labourers carried earth up, basket-load by basket-load. Then, when it had bedded down, native and exotic species were planted out.

Opened in 1864, the Botanical Gardens were soon known as *Ping Tau Fa Yuen*—or Head Soldier's Flower Garden, for their proximity to Government House. The Gardens proved a blessing in a town where access to countryside was limited by the rugged terrain. Moreover, in a community where Chinese–European antipathy was pervasive and often virulent, and where Chinese were banned from the City Museum and Library, the Botanical Gardens were open to all Hong Kong people. East and West mingled there, albeit superficially, under the welcome shade of both Chinese and European trees.

By the mid-1870s the Botanical Gardens were well established. In 1878, as one woman visitor wrote, 'rich and rare forms of foliage from tropical and temperate climes combine to produce a garden of delight'. And when that intrepid traveller, Miss Isabella Bird, passed through Hong Kong in the same year, near the Botanical Gardens she was impressed by balconies 'festooned with creepers' and 'grand flights of stairs arched over by dense foliaged trees'.

As early as the mid-1840s avenues of trees had been planted around Victoria, and by the 1860s they gave much-needed summer shade. After his visit in 1869 the photographer John Thompson wrote: 'A splendid town has been built out of the barren rocks, and the hill-sides are covered with trees'.

As Thompson's own photographs show, in fact government tree planting was limited to the slopes immediately above Victoria. In the late 1860s the higher slopes were still rocky and treeless, and photographs of the hills above Happy Valley show mostly barren slopes. The Kowloon hills were totally bare. Given the sudden growth of a poor urban population, most of Hong Kong Island was almost certainly even more denuded then than in 1841 because of people combing the hills for fuel. Small mammals and birds must also have been depleted, as the 1870 *Preservation of Birds Ordinance* suggests.

More widespread reforestation of Hong Kong Island began in the 1870s. The result of Western botanical science, British colonial administration and Chinese energy, the reforestation was ambitious—and most successful.

The Gardens and Afforestation Department was established by Governor Sir Arthur Kennedy (1872–77), whose enthusiasm for public works was matched by ample funds. The driving force behind the early reforestation was Charles Ford, a young botanist who had come to Hong Kong in 1871 as Superintendent of the Botanical Gardens. As G. A. C. Herklots wrote:

> *In 1872 [Ford] began experiments planting trees on the bare hills. Being satisfied with the results, in 1876 he advised the Governor to inaugurate and organise a system of afforestation adapted to Hong Kong, so that the comparatively treeless island might be redeemed.* *

The scale of reforestation grew dramatically during the governorship of Sir John Pope Hennessy (1877–82), who also doubled the size of the Botanical Gardens. Almost all the plantation trees were Chinese Pines—since the species could survive even on exposed sites—and the numbers planted by Ford's Chinese labourers were remarkable. In 1884, for example, 714,159 trees were planted, 700,000 of them pines and the rest various native and exotic species.

The 1888 departmental report stated that 682,325 trees had been planted out that year, and gave a total of 5,676,207 planted since 1876. However, various factors constantly worked against the reforestation: poor shallow soils; summer typhoons and rainstorms; winter hill fires; and (now illegal) tree-cutting. As Governor Des Voeux (1887–91) noted against the 1888 totals, many of the seedlings and saplings probably failed.

* *As the historian E. J. Eitel noted, Kennedy 'was one of those few men who deserve a statue because they do not need one'. In fact a statue commemorating the far-sighted Kennedy was placed in the Botanical Gardens, but it was destroyed during the Second World War. Ford's excellent botanical library was also destroyed during the Japanese occupation; but his irreplaceable herbarium, sent for safekeeping to Singapore in 1941 before the expected invasion, miraculously survived the war.*

Yet, by dint of experience and sheer toil, the Hong Kong hills grew steadily greener. What the foresters learned was something local villagers had known for generations: that, despite the poor soil on Hong Kong's hills, the hot wet summers bring on luxuriant growth if plants are left alone. Thus, when the Revd J. A. Turner visited in the late 1880s he wrote that 'the island of Hong Kong is now well wooded'. He continued:

> *The whole place is a paradise of semi-tropical beauty, and the greatest possible contrast to the bare and uninteresting aspect of the neighbouring mainland.*

In 1841 Lord Palmerston had famously dismissed Hong Kong as 'a barren island with hardly a house upon it'. By the 1870s and 1880s, there were far too many 'houses'—and people. The 1876 and 1881 censuses recorded populations of 139,144 and 160,402 respectively. Thus, in just five years, the population *growth* was about five times the *total* pre-1841 land population. But the rainfall, and the flow of water off the hills, were virtually the same as before.

Hong Kong's monsoon rain almost all falls in the summer months, and the terrain provides no natural water storages. Villagers had always relied for water on streams in summer and wells in winter. As the urban population grew, more wells were sunk, many through the foundations of buildings. But by the 1860s even they were inadequate—and increasingly polluted.

Reservoirs became the inevitable, costly solution. The southern side of Hong Kong Island before its reservoirs were built was as it had been for generations. Its indented bays and deep valleys had just a few Hakka and Tanka farming-fishing villages, linked by steep mountain pathways. Now, driven by the need for water, urban Hong Kong extended its reach across the central peaks to the Island's southern side—at Pok Fu Lam, Tai Tam Tuk, Aberdeen, and later Wong Nai Chung.

The first Hong Kong Island reservoir, at Pok Fu Lam, was completed between 1863 and 1864; and the last, at Tai Tam Tuk, was opened in 1917 or 1918. The early reservoirs, impounded behind dams built across narrow upper valleys, allowed water to flow by gravity around the Island to Victoria. But the later dams enclosed wide lower valleys, submerged villages and fields, and needed major tunnels and pumping systems to take water through the mountains to the city.*

The Hong Kong Island reservoirs were, for their time, massive undertakings. The population's inexorable growth demanded the

* *The precise completion dates for some reservoirs are uncertain, due to the loss of records during the Second World War.*

Rapid population growth, combined with the lack of river and lakes, demanded reservoirs to provide an urban water supply. Tai Tam Reservoir, shown here, was built during the late 1880s.

building of a new reservoir every decade. The reservoirs grew steadily larger: Tai Tam Reservoir, completed in the late 1880s, held ten times the volume of Pok Fu Lam Reservoir, completed some twenty years before; but Tai Tam Reservoir was itself one quarter the size of Tai Tam Tuk Reservoir, completed some thirty years later. Around all the reservoirs reforestation was begun: to even out the water drainage; to minimize erosion; and so to prevent the reservoirs filling with silt.

The catchment channels also provided walking trails, but hiking over the Island to get to them was taxing. It was far easier to take a ferry to Kowloon, as the *Hong Kong Telegraph* suggested on 8 October 1897. With the cooler weather, the newspaper noted, exploring beyond 'Yau-ma-ti' village made a pleasant afternoon walk—especially as the second rice crop was 'already in the ear':

> *When the collecting ground of the waterworks is passed, the road winds up over bare and rugged hills, which are a great contrast to the fertile valleys below. Plunging down again, as it skirts the Chinese cemetery, [the road] leads on past more paddy fields and gardens, until the bamboo boundary fence, dividing British from Chinese territory, is reached. Along the boundary fence, in the paddy fields, snipe and an occasional wild duck are to be seen.*

Hong Kong's climate, as the author of *English Life in China* put it, was 'alternately unquestionably healthy and deplorably sickly' depending on the season. In the summers, as the photographer John Thompson remarked, 'Books and papers become limp and mouldy, and the residents feel as in a vapour bath, reclining in their chairs and languidly watching the flying ants.' When Prince Henry of Prussia visited the colony in the summer of 1880, Governor Sir John Pope Hennessy wrote to Queen Victoria's lady-in-waiting:

> *I have begged him to avoid going out in the middle of the day, and never to do so, when the sun is shining, without his sun-hat and umbrella.*

Towards the end of the nineteenth century some developments made the summers more bearable, especially for Europeans. Ice had been imported since 1845, and by the mid-1870s some was manufactured locally. Summer meals became healthier and more varied when ships began carrying frozen meat in the 1880s. For a fortunate few, there was access to the Peak's cooler temperatures. And, for those unable to endure another Hong Kong summer, after 1869 the Suez Canal meant that people could more quickly—and cheaply—go 'home'.

The Royal Observatory, sited on a hill in Tsim Sha Tsui, began recording the weather on 1 January 1884. Part of an expanding world-wide network of meteorological stations, it brought new objectivity (and continuity) to measuring and predicting Hong Kong's weather: the mean annual rainfall was about 2,200 mm, the temperature ranged over about 10°C through the year, and the summer humidity was often over 80 per cent. The summer monsoon was still loathed and its departure welcomed with relief. And meteorology could do nothing to prevent or control nature's outbursts: the summer monsoon's typhoons and rainstorms.

Rarely did a Hong Kong summer end without at least one typhoon having threatened the place. Indeed, as G. R. Sayer wrote, 'Typhoons began to weave themselves into the texture of Hong Kong life.' Severe typhoons that came close enough to cause major destruction were unusual but, with Hong Kong's large population, the potential for major loss of life was greater than before. In October 1867 there was a violent typhoon; and in late September 1874 an even more devastating one lashed Hong Kong. Some 2,000 people were killed, thirty-five ships were driven ashore and countless junks and sampans were sunk. As the *Hong Kong Times* reported:

> *The wind blew with the violence of a tempest, the rage of a whirlwind. Vessels staunch and strong were driven hither and thither about the harbour or on to the shore.... Trees were uprooted by the hundred; rows of buildings were blown down in a moment, many of the inhabitants being buried beneath the ruins.... Among other calamities the praya wall, laboriously rebuilt after its destruction in 1867, was once more demolished.*

Photographs show that during the 1880s and 1890s the Peak was still rocky and virtually treeless. Exposed to the full force of wind and rain, its sheer slopes eroded easily. Victoria, by contrast, was far better sheltered. However, its streets were buried under tonnes of mud and rock after intense rain—such as the deluge that lasted from 28 to 29 May 1889. In just thirty-three hours some 900 mm of rain fell, as Governor Des Voeux recorded in Mountain Lodge:

> *For two whole nights and the greater part of the intervening day the thunder and lightning was almost incessant. Sometimes for an hour or more flashes succeeded one another so rapidly that even in the middle of the night it would have been possible to read by them.*

Near Mountain Lodge, some Chinese labourers were caught by the storm in a flimsy shack. They must have been petrified. Then their ordeal was over: a brilliant flash pierced the sky—and the eight men were transfixed by lighting. Only a few other people were killed, but the physical destruction was immense. As the Revd Turner noted, the total rainfall was greater than Britain's annual average. Over Victoria, he estimated '230,000 tons [of rain] fell per hour':

> *Such a terrific deluge naturally dislodged enormous quantities of earth and stones.... The main drains of the city burst, and roads were torn up and converted into deep ravines. All the streets became rivers, along which water, bearing earth and stones, rushed with terrific force.... Never will those who witnessed the havoc forget it, or cease to reflect on the awful forces of nature.*

Winters were generally a time of mild, pleasant weather. But for a few exceptional days in mid-January 1893 icy winds surged down from North China. The temperature at the Royal Observatory dropped to freezing, and on the Peak to –4°C. The mountains turned white with hoar-frost and icicles formed in the rigging of ships. Tropical plants were severely affected, and many of the Botanical Gardens' orchids and ferns died.

Hong Kong's extreme terrain and climate demanded human energy and resilience. The challenging topography nurtured a spirit of bold enterprise and demanded ambitious projects, which created employment, and so attracted more people. By 1898 the population was about 250,000, including 15,000 Europeans. By then the colony was one of the world's largest and busiest ports.

The New Territories were leased from China in 1898, but about a decade passed before their valleys and uplands were effectively integrated into the Colony. On Hong Kong Island, by contrast, still greater environmental changes were heralded by three developments: the introduction of electric power, the first major harbour reclamation, and roads reaching into the countryside.

The Hongkong Electric Company began generating power on 1 December 1890. The dry winter weather was an advantage, as rain might have extinguished the first fifty electric-arc street lamps. However, when the company's Wan Chai coal-fired generator developed faults in its steam condensers, more water was urgently needed. It being winter, the Wan Chai nullah had dried to a trickle. So Hakka women, with poles and buckets, were employed to scour Hong Kong Island for water.*

Victoria's shortage of land had always been limiting. Some small reclamations had extended the foreshore out marginally, and so destroyed much of the harbour's natural charm. But the Central Reclamation, begun in 1890, was the first major scheme—beginning a process that continues today. It created about 23 hectares of land along a three-kilometre strip and took 14 years to finish. The reclamation was an immense achievement: for, basket-load by basket-load, Chinese labourers—men and women—unloaded its fill from junks and sampans.

Between 1885 and 1888 the construction of the Peak Tramway overcame extremely challenging terrain to provide far easier access to the Peak. A decade later, to mark Queen Victoria's Diamond Jubilee in 1897, the government announced plans for ambitious roads which would connect Victoria to the Island's southern bays. The news was greeted with enthusiasm. For, as views from the Peak showed, the Island's southern side was majestic—and, except for its reservoirs, virtually untouched.

In the six decades since 1841 Hong Kong Island's half-wild, half-agricultural landscape had been transformed. And it was not surprising that, when an 1897 Singapore *Straits Times* report commended 'Hong Kong's progress', it was reprinted here in the *Hong Kong Telegraph* on 5 January 1897. As the report observed, the Colony was 'being transfigured':

> *No place in the East is changing as rapidly as Hong Kong. The tramway to the Peak, the extending reclamations and other factors, are making Hong Kong a city which may at no distant date vie in magnificence with Calcutta—but we hope it will be kept cleaner.*

Transforming Hong Kong's natural landscape demanded massive human effort.

* *For some time many Chinese shied away from using electricity because of its mysterious invisible 'spirits'. More than two generations passed before electricity was widely used by poorer Chinese, who until then continued cooking with wood and charcoal.*

Tai Tam Tuk Reservoir, seen from below Mount Parker.

REGION

HONG KONG ISLAND

Climb with me ... some fourteen hundred feet up to the gap by the Peak, and find language to describe the scene of cliff and scar and winding bay, of hill and vale, mountain and island and rock visible on either hand; for here we stand on the very backbone of our island.

SYDNEY SKERTCHLY, 1893

Below my bivouac, hillsides plunge to a lush wooded valley, a natural amphitheatre set amid peaks and ridges. The hills, in places scattered with boulders, are mostly emerald green.

It is early summer. Dusk is drawing a veil over the Tai Tam valley and its reservoirs. As the shadows lengthen the afternoon breeze drops to a zephyr. Jade green, the rippled reservoirs flatten to a mirror sheen. For a minute their red-earthed promontories, like the jutting prows of galleys, glow richly. The only sounds are of birds nesting, and water gurgling down a gully.

Far away, framed between the hillsides, Lamma, Po Toi, and Beaufort islands rise abruptly from the sea. Sunset brings a dramatic cloud display, a kaleidoscope of swirling, changing colours: greys, yellows, oranges, crimsons.

And this is Hong Kong Island! I am camped along one of the Tai Tam reservoir catchment trails. There is no one nearby. Yet, visible just two kilometres away, are the homes of some three thousand people—residents of the massive 'Parkview Suites' complex. More remarkably still, just beyond the ridges above where I am, is the Island's densely urbanized, overcrowded harbour fringe.

Hong Kong Island, despite its fine harbour, was in the nineteenth century far from hospitable. Barren slopes fell to mere scraps of foreshore and the few valleys were swampy. Penetrating its rugged interior demanded exhausting effort, especially in summer. The Island had changed little half a century later, when the botanist Sydney Skertchly was here. Indeed, as late as 1920 most of the southern Island still had no proper roads.

Today spectacular highways encircle Hong Kong Island, and walking trails cut across its rugged uplands. Here, more than anywhere else in the Territory, city and country meet face to face: tower blocks stare out at forested hillsides, ridge-top paths look down on freeway interchanges.

Fortunately much of Hong Kong Island was too steep to be easily developed. Indeed, from the late nineteenth century much of it was effectively conserved as countryside, when valleys were turned into reservoirs and their surrounding hills reforested. Today most of the Island above about 200 metres—40 per cent of its total area—lies in four Country Parks. The smallest, Pok Fu Lam, has just 270 hectares; the largest, Tai Tam Country Park, covers over 1,300 hectares.

Hong Kong Island's area is 78 square kilometres, and it extends for about 14 kilometres from west to east. Yet the Hong Kong Trail, a walking track linking the four Country Parks, winds through dramatic country for about 50 kilometres. Traffic often strangles the urban areas, but the trail crosses just four public roads in its entire length.

Queen's Road East, Wan Chai (Little Bay): surely just another urban canyon? But look closer. Here Hong Kong Island once met the harbour—before reclamations pushed the foreshore 700 metres further out. The Hung Shing Temple, probably completed in 1860, is built literally against old coastal boulders. Behind the temple nature still keeps a tenuous hold. An Indian Rubber Tree shades the temple, its roots and tendrils encasing a slab of rock. The immediate surroundings are all concrete, but the tree survives off nutrients and water that seep underground from the hills.

From near Stone Nullah Lane the Wan Chai Green Trail winds steeply up to Wan Chai Gap, following the old village path over the spine of Hong Kong Island. Just up the trail, ferns and shrubs grow

Dense vegetation near Wan Chai Gap. Only decades ago, after the Second World War, the same area was severely denuded.

from old stone retaining walls. Further up Flame of the Forests and Jacarandas spread feathery boughs above the path, and water runs down gullies dark with mosses. Earth shrines line the path. Higher still, the slopes are thickly wooded.

A booklet on the trail notes with striking understatement: 'The change from a highly urban environment to the woodlands on the hill slopes is particularly worth noting.' It surely is. By 1945 Hong Kong Island had been severely deforested by wartime tree-cutting for fuel. But today, as the Wan Chai Green Trail shows, the woodland has grown back—except where the hills are too sheer and rocky for soil to form, let alone plants take root.*

Wan Chai Gap, at almost 300 metres, is the most central pass crossing the Island. To north and south are dramatically different vistas. There is still the great natural beauty that Sydney Skertchly saw, 'cliff and scar and winding bay'. But now there is also the awesome grandeur of man-made Hong Kong.*

To the south, the coast is deeply indented with coves and bays, bluffs and headlands. To the north, the shoreline was originally also sinuous though much less jagged than in the south. But today straight-edged reclamations line the entire north of Hong Kong Island. The tramline that runs from west to east, from Kennedy Town to Shau Kei Wan, lies close to the original shore; and Possession Point, where the British raised their flag over Hong Kong, is now stranded 500 metres inland.

Wan Chai Gap brings home another contrast. Beyond Aberdeen valley's forested slopes, the air is at least relatively clear, and the sea, though polluted, is still usually blue-green. But around the harbour a pall of pollution generally hangs over the urban areas.

* *The* Wan Chai Green Trail Booklet *is by Dr C. Y. Jim, known for his research into Hong Kong trees in both rural and urban settings. The booklet is available at the Environmental Resource Centre in Wan Chai.*

* *The quotation from Sydney Skertchly is taken from a general reference book. I was unable to locate Skertchly's own book,* Our Island a Naturalist's Description of Hong Kong *(1893). Given its year of publication, some twenty years after widespread reforestation began, it may well contain observations concerning the progress of reforestation and any resulting increases in fauna habitats and populations.*

A symphony of peaks. The southern side of Hong Kong Island is extremely rugged, as this view from Lamma suggests.

The harbour itself, fouled with the filth of decades, is often more grey than blue-green. One wonders whether keen-nosed animals can smell the harbour from Wan Chai Gap.

Down in Wan Chai itself mosses and ferns grow. But lichens, primitive plants that are extremely sensitive to air pollution, no longer can. However, in the hills above Wan Chai Gap the air is cleaner, and lichens can still be seen.

East of Wan Chai Gap the hills above Black's Link are thickly vegetated—and uninhabited. After rain crystal clear water seeps from springs along the path. Below it, watercourses plummet over sheer drops, disappear into impenetrable gullies, and finally run down to Happy Valley.

Black's Link looks directly over Happy Valley. In 1841 the valley, the Island's only significant flat land, was covered with paddy fields and swamps. The British flattened the nearby hillocks, reclaimed the swamps, and built granite residences. Then disease struck: it decimated Hong Kong's Europeans, especially those living in the 'happy valley'. So Happy Valley's residences were abandoned, and tropical growth invaded their gardens.*

Glimpsed from Black's Link, lying against Happy Valley's lower slopes, are the Colonial, Catholic, Parsee, and Jewish cemeteries. Each is a poignant reminder of how harsh Hong Kong Island's environment once was.

It was an oppressive summer afternoon when I wandered through the Colonial Cemetery. Between streams that descend from Black's Link, among boulders and buttressed trees, pathways wind among the dead. The individual sadnesses are touching: husbands, wives, and children swept away by tropical afflictions. The wider picture is appalling: mortality on a massive, tragic scale.

One tall plinth states: 'Sacred to the memory of the non-commissioned officers and men of the 95th Regiment, who died at Hong Kong in the summer of 1848, this column is erected by their comrades. Died of fever 1st June–30th Sept, 9 sergeants, 8 corporals, 4 drummers, 67 privates, 4 women, 4 children.'

A tablet gives the epitaph for half a ship's crew: 'They shall hunger no more, neither thirst any more, neither shall the sun light on them, nor any heat.'

Unsung and unrecorded, countless poor Chinese also died during those years. The people whose toil literally—*physically*—reshaped Hong Kong Island: the stonemasons, the brick carters, the sand carriers, the boat people, and many more.

* *Lantana, introduced to Hong Kong as a garden plant, was first grown here in Happy Valley in the 1850s. It spread rapidly, and today its multi-coloured flowers are seen virtually everywhere.*

Black's Link ends at Middle Gap, where Mount Nicholson rises abruptly along the Island's central spine. One crosses the divide quickly: within moments Happy Valley's cemeteries and massed buildings give way to deep green hills rolling down to Wong Chuk Hang (Yellow Bamboo Creek) near Aberdeen. Bays and islands stretch into the distance.

The trail running eastwards skirts Mount Nicholson's southern flank. It is late autumn, and reptiles are busy foraging before hibernating: a Blue-tailed Skink slithers across the path, its tail glistening aquamarine. Snakes mostly keep away from paths, but among the forty-odd local species there are a few venomous land snakes: two mildly venomous back-fanged species, and more dangerous kraits, vipers, and cobras—like the 'dark object close to my feet' that the photographer John Thompson almost trod on near Wong Nai Chung in 1868:

> *I raised my camera in order to use the tripod as a weapon of defence, whereupon the reptile reared its head, erected its hood, and with a hiss slid down off the rock into the underwood.*

Leaves in a watercourse near Wong Nai Chung Gap.

Further round is Wong Nai Chung Gap (the name means Yellow Muddy Stream Gap, hinting at previous deforestation and erosion). Nearby are the 'Parkview Suites', built amid some of the Island's loveliest hills. Parkview's sheer bulk totally dominates its setting. The complex is a bleak comment on the greed of some Hong Kong developers, whose only concern is the height of their building. Parkview was developed on a deep horse-shoe bend of land, originally excluded from Tai Tam Country Park; the Country Park surrounds the complex on three sides.

Above Wong Nai Chung Gap is Violet Hill. Not since I was a teenager had I climbed this way, along the airy ridgetop path that ends high over Tai Tam valley and Repulse Bay. Wind-bent pines and stunted bamboos line the way, with tantalizing gaps between them revealing plunging valleys and wooded hilltops. Half an hour up from Wong Nai Chung some clay steps wind up towards the summit of Violet Hill.

Even for Hong Kong, replete with panoramas, Violet Hill is overpowering. Below its 433-metre summit, slopes sweep down into the depths of Tai Tam valley, past the lizard-shaped Tai Tam Intermediate Reservoir, on to the lake-like expanse of Tai Tam Tuk Reservoir. It is late afternoon: the dam walls reflect brilliant sunlight, their earth banks red-brown around aquamarine waters. The forested hillsides are cut by gullies that cast dark shadows. Behind me, seen through tall tufted grasses, framed by the sides of Wong Nai Chung Gap, is urban Hong Kong's dramatic skyline. The sun glints on skyscrapers, and the Bank of China Tower throws up gleaming triangles amid the grasses.

To the left, a roller-coaster ridgeline runs from Jardine's Lookout, past Mount Butler and on to Mount Parker. Beyond there is the Dragon's Back peninsula. To the right, Violet Hill's sheer slope blocks the view. But further along the track Repulse Bay's sublime setting appears: Middle Bay and South Bay, with their humpbacked islets; Round Island; and the bluff boulder-strewn Chung Hom Kok promontory. Further round, Lamma Island, Cheung Chau, and Lantau float dreamlike across a hazy pale blue sea.

The Parkview complex dominates Tai Tam valley's serene setting. This intrusive development was built on land originally excluded from—but virtually encircled by—Tai Tam Country Park.

G. S. P. Heywood, a keen hill walker in the 1930s, wrote of Hong Kong Island's 'serene' countryside: 'Bustle and unease are forgotten, and it seems incredible that a couple of million people are pursuing their business a few miles away beyond the hills.' The proximity of city and country are still more remarkable today.

Gazing down from Violet Hill into Tai Tam, one is looking over both the ancestral domains of Chinese clans, and the engineering feats of British colonial Hong Kong. As James Hayes says there is 'an epic quality' about Tai Tam valley: 'Its scenery and the fate of its village are equally compelling'. It is sobering that Parkview's 3,000-odd people equal at least half of Hong Kong Island's total population of villagers and boat people in 1841.

The south side of the Island had too little flat land and too few good streams for extensive agriculture. But in the Tai Tam valley—the name means 'edge of the big waterway'—two Hakka clans had established paddy terraces after their ancestors settled there in the late seventeenth century. Their brick homes stood near the valley mouth, as James Hayes has established.

Eighty-odd Hakkas and some Tanka boat people lived there in the 1840s. Robert Fortune, exploring for plants around Tai Tam in 1844, 'found the inhabitants harmless and civil'. A decade later an artist, William Wynne Lodder, captured 'Tytam Took' on canvas: a sylvan valley, a wide stream, a wooden bridge, a cluster of houses, a *fung shui* wood (the work is in the Museum of Art). In 1858 an *Illustrated London News* correspondent, almost certainly describing Tai Tam, reported:

> *On the 1st of November [1857] I walked into the interior of Hong Kong, and saw the process of rice harvesting. Beneath a bright, hot sun, the entire village population was hard at work getting in the second crop of paddy. The principal part of the labourers are women, owing probably to the fact of the men being generally engaged in fishing.*

The three Tai Tam reservoirs were triumphs of engineering. Tai Tam Tuk, the largest of them, demanded this lengthy dam across the valley's mouth.

From Violet Hill a path winds along a steep spur. Halfway down to sea level, about 200 metres above Repulse Bay, it turns inland—and drops into the Tai Tam valley, past Tai Tam Intermediate Reservoir.

The construction of the three Tai Tam reservoirs reflected urban Hong Kong's urgent need for more water—and spelt doom for the Hakka villagers' domains. Hong Kong's population was already over 180,000 when Tai Tam Reservoir, the first Tai Tam water scheme, was completed in the late 1880s. When Tai Tam Tuk Reservoir (which submerged the Hakka village) was completed between 1917 and 1918, the Colony's population was well above 400,000—and would have been over 500,000 but for the First World War. Despite the villagers' personal and spiritual objections, their lands were flooded. The same streams that had brought down water and alluvium, that had enriched their fields, now made the valley ideal for impounding water to supply the city.

The Tai Tam Reservoir scheme was far-sighted, ambitious, and very costly. Governor Sir William Des Voeux visited the site on 23 November 1888. A keen shooter, pleased to be out in the country, Des Voeux travelled 'over the backbone of the Island'. As he wrote:

> *The journey over and along the hills was most enjoyable, and, as we used our chairs whenever inclined, very luxurious. Occasionally we went down in a valley, which might have been in the middle of a Scotch 'forest', so devoid was it of any sign of human existence.*

The view into Tai Tam valley revealed a different scene: hillsides scarred red-brown with the reservoir workings and prodigious, ant-like industry: 'some two thousand coolies at work, several steam engines palpitating, and all the sounds and signs of a big undertaking', Des Voeux wrote.

Stand on the dam wall of Tai Tam Reservoir today and one can imagine what Des Voeux saw in 1888. About two kilometres above Tai Tam Bay the upper valley had already been enclosed behind a dam. The retaining wall, constructed with Hong Kong granite and Portland cement, was 30 metres high, 20 metres thick at the base, and topped with a small turret. A tunnel under the hills, to take water to Victoria, was being driven through granite bedrock.

The completed Tai Tam Reservoir held about 1.4 million cubic metres of water. Yet within a year Hong Kong was brought close to collapse when the monsoon rains failed abysmally. Another drought in 1902 preceded the construction of Tai Tam Intermediate Reservoir, completed between 1907 and 1908 and holding just under a million cubic metres. More erratic seasons—and more people—finally led to Tai Tam Tuk Reservoir being built. Completed between 1917 and 1918, it held 6.5 million cubic metres, then an enormous reservoir volume.*

The engineering achievements of these reservoirs, and their role in helping Hong Kong prosper, were not lost on those wishing to improve conditions in China. One such patriot was Pan Fei-shing—a newspaper editor, poet, and painter—who arrived in Hong Kong in 1892. Sometime later he explored around Tai Tam, as his essay 'Roam About Tai Tam Tuk' describes:

> *The reservoir is a few hundred feet wide and meanders among some ten hills.... The sources of the reservoir [are] just many small pools into which water trickles, but without the Englishman's surveying and planning, without the steam engine, nothing like this would ever have been achieved.*

Bitterly conscious of the poverty in his Guangdong village, not a day's journey away, Pan added: 'If the nature of the land is not studied, if man does not exert himself, even fertile fields will become barren and useless.'

Besides their water, the Tai Tam reservoirs, and the smaller Hong Kong Island dams above Pok Fu Lam and Aberdeen, brought two great benefits to the Island's countryside: water catchment channels and reforestation. Catchment channels reach like tentacles through the Island. Encircling the hills, they channel water run-off during—and usually well after—the summer rains. The Agriculture and Fisheries Department quaintly describes catchment areas as 'water gathering grounds'.

When they were built, the catchment channels were major feats of surveying and excavation. Around entire hills channels were cut that descended evenly through a few hundred metres. All were concreted. Some, like those above Tai Tam Reservoir, were lined with ceramic tiles. Today their hillside cuttings are camouflaged under mosses and ferns. But the drill marks are still there—and jagged rock where dynamite blasted the hills apart. How many labourers were killed building the reservoirs and catchments, one wonders?

The catchment channels offered another advantage. Across slopes often too rough for pleasant walking—sometimes too precipitous even for scrambling—they provided invaluable hiking trails. As the hills became steadily thicker with trees, the

* *Due to the destruction of waterworks records during the Second World War some dates concerning these reservoirs are in doubt. Hence, as existing sources conflict, these completion dates are given as 1907 to 1908 and 1917 to 1918. The Hong Kong Island reservoir structures are some of the best preserved, and least threatened, parts of Hong Kong's 'built' heritage.*

channels' access value increased. On the harbour side of Hong Kong Island, the conduit that took Tai Tam's water to the filter beds above the Botanical Garden provided a scenic promenade along the hillsides.

Reforestation of the surrounding hills went hand-in-hand with each new reservoir. Three natural factors assisted the shrubs and trees planted around Tai Tam to establish and later to self-propagate. The area was well sheltered; the lower valley had rich volcanic soils (like Tai Po Kau's); and the rainfall was high. The distance from Victoria also meant there was less illicit wood-cutting than elsewhere by urban dwellers. As a result, Heywood could observe in the 1930s: 'The southern side of the Island does not seem to have suffered much from the ravages of woodcutters, and most of the lower hillsides are well covered with trees or scrub.'

The Second World War changed that. Hong Kong Island was savagely fought over during December 1941. Bombs, shells, and fires severely damaged the dry winter hills. But what followed was far worse. During the Japanese occupation Hong Kong people, desperate for fuel, took to the hills—and cut down the vegetation established so laboriously during the past decades. The toll on shrubs, trees, and topsoil was devastating.

Climbing above Repulse Bay and Tai Tam during the late 1950s, my friends and I fought boyhood 'battles' amid the evidence of real war: pock-marked hillsides; low, thin vegetation; and wartime 'pillboxes' still scattered with spent cartridges. That the same pillboxes high above Tai Tam are today virtually buried under dense vegetation attests to the success of government reforestation since 1945.

Today, throughout Hong Kong Island, the Country Parks' hills and valleys are covered in shrubs and trees. The lower valleys have thick woodland entwined with lianas. Indeed, off pathways the vegetation is generally far too dense for people to penetrate. Steadily developing into 'mature' woodland, most of the Island's forests include native broad-leaved trees, pines, and introduced species. Hill fires are less common on the Island than elsewhere, partly because there are few hillside graves.

The reforested hills, by absorbing more rainfall, greatly even the water run-off after summer rains. Both reservoirs and reforestation have created invaluable habitats. Frogs, terrapins, and water snakes live in the catchments and reservoirs; and the reforested hills have been recolonized by some mammals and woodland birds. Small numbers of Barking Deer, Pangolins, Chinese Porcupines, Chinese Ferret Badgers, and Civet Cats still live on the Island.

D'Aguilar Peninsula extends southwards for five kilometres from the head of Tai Tam Bay. Along much of it runs an up-and-down ridge, the Dragon's Back (Lung Chek).

The Europeans who first visited the peninsula described it as windswept and without even shrubs. Despite some regeneration, the higher peninsula is still dominated by grasses and low shrubs. Unlike Tai Tam the ground is granitic, and so weathers and erodes easily. The land takes the brunt of summer typhoons and winter winds, which reduce the ground cover and so increase soil loss. And rainfall is lower than further inland. Together, these ecological controls have prevented woodland from developing, except in some deep sheltered areas.

D'Aguilar Peninsula's coastline more than compensates. Hiking along the Dragon's Back—and down to the sea—brings one closer to Hong Kong Island's ocean face than almost anywhere else. From Cape Collinson to Cape D'Aguilar (Hak Kok Tau to Hok Tsui) is six kilometres in a straight line, but ten or more measured around the indented coast.

Immediately to the north of Cape Collinson reclamation has totally obliterated the natural coastline. But to the south towering cliffs drop into the water; ocean swells smash onto the land; and sea caves hint of monsters. Off Shek O the island Ng Fan Chau is rocky and wild. And so, past plunging slopes, one arrives at Cape D'Aguilar—Hong Kong Island's delightful Land's End.

Over fifty bird species have been recorded there. The cape, protruding into the South China Sea, is a beacon for migrating

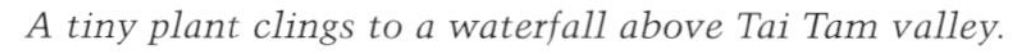
A tiny plant clings to a waterfall above Tai Tam valley.

Government reforestation has transformed Hong Kong Island. In the 1950s, this fortification above Tai Tam was easily accessible; now it is enclosed by tall shrubs.

birds and is often their first local landfall—and in typhoons offers respite for storm-driven wanderers.

The old Cape D'Aguilar lighthouse stands on a hill. Below it, the Swire Institute of Marine Science, the most south-easterly building on Hong Kong Island, sits just behind the cape. Recently expanded, the Institute conducts research into Hong Kong's marine ecology. It is urgent work—as much of it has to do with the degradation of Hong Kong's waters. Indeed, even here near the ocean, the sea is polluted—and silted from dredging in the nearby sea channels. Beside the Institute the skeleton of a Common Rorqual whale stands on a plinth: the bleached remains of the creature that became ensnared beneath a harbour pier in April 1955.

Cape D'Aguilar is one of Hong Kong Island's loveliest—and wildest—places. During tropical storms the ocean rages, washing over its projecting rocks with seething swells; and during typhoons the jagged coast is smothered under monstrous waves and flying spume.

At other times the seascape is serene and tranquil. The last time I was there, minutes before dawn the sea was flecked with pink and gold. As the sun rose, glistening water extended like a golden river to the eastern horizon.

Cape D'Aguilar's actual tip, a castle-like rock, glowed yellow, then orange, then gold. A sea cavern, flanked by rock pinnacles, swooshed with each incoming swell. Across Tathong Channel, peak rose atop peak. Out to sea stood steep-sided islands, remote and desolate: the Ninepins, dark wedges of rock; Waglan, its light still faintly flashing; Beaufort and Po Toi islands, forbidding and splendid.

Could Wan Chai's teeming streets really be just twelve kilometres across the hills?

Cape D'Aguilar soon after dawn. Beyond the nearby islet, Po Toi and Beaufort Island rise abruptly three kilometres out to sea.

Cape D'Aguilar, the south-eastern tip of Hong Kong Island, is extremely exposed and takes the full force of storms and typhoons.

By the 1930s the Hong Kong countryside's remote parts were fairly accessible, but 'progress' had so far barely altered its agriculture and ecological balance.

HISTORY 1900–1941

ENDURING DOMAINS

I began to wonder how it was that foreigners, that Englishmen, could [achieve] such things as they have done, for example with the barren rock of Hong Kong.

SUN YATSEN, 1923

The hills above Victoria are clad with a thick carpeting of trees, and a shimmering mist tones the harshness of the mountains and the buildings. Sometimes there are days of intense calmness, with sunshine breaking through the clouds. Then Hong Kong has the soft loveliness of a Japanese print, and Tai Mo Shan is a smaller Fujiyama.

IN FAR EASTERN WATERS, 1930S

On a bright winter day in December 1922, a government cadet stood spellbound as his ship steamed into Hong Kong harbour. He had read about the Colony, but nothing had prepared him for its natural setting—or its energy. Yet the harbour's ceaseless activity seemed to blend with the landscape: 'nature and man not in conflict but complementing one another', he wrote.

Alexander Grantham, the young cadet and a future governor, immediately recognized Hong Kong's distinctive feature: the close proximity of its natural and man-made grandeur. Later that day he strolled past banyan- and fern-covered slopes up to the Peak Tramway. From the Peak itself he looked out over well-wooded hills: across the harbour and towards the South China Sea, where islands rose 'from a glass-like sea in a crimson sunset'.

'Nature' indeed lay close to the harbour and Victoria, though it could be far from serene. Sixteen years before Grantham arrived the catastrophic typhoon of 1906 had blown in off his 'glass-like sea'.

On Tuesday 18 September 1906 dawn broke dull and threatening. The barometer had been falling, indicating that a typhoon might be approaching. However the Observatory, unable to confirm one, forecast 'variable winds, moderate, probably some thunder showers'. In fact, just out to sea, a typhoon was suddenly intensifying. Across Hong Kong the air pressure suddenly fell alarmingly. Then, with virtually no warning, gale force winds raged in.

The typhoon passed directly over the harbour and, in just two hours, devastated Hong Kong. Forty-one ships were driven ashore, and over 2,400 junks sunk. The usually bustling harbour Praya was strewn with wreckage, and the stench of death hung there for days. Probably about 11,000 people were killed. But, as one newspaper mourned: 'The roll of death will never be known. In most cases whole families have disappeared together and there is none to ask for them.'

Two fundamental facts affected Hong Kong's natural setting between 1900 and 1941. First, following the leasing of the New Territories in 1898, the physical landscape was greatly extended. Second, because of events in China, refugees swarmed into Hong Kong and severely affected its urban and rural environments.

By 1900 Hong Kong's heart, Victoria and its environs, was largely urban; Kowloon was soon to be thus transformed. The New Territories, by contrast, were totally rural. There fields and villages covered the fertile lowlands, while the uplands were half-agricultural, half-wild.

The 1920s and early 1930s were golden years for the Hong Kong countryside. The New Territories' mainland, and their islands, covered about twelve times the area of Hong Kong Island, and had far more varied country. New roads and a railway brought vastly improved access into the hinterland; more reservoirs were built; and reforestation continued. Yet the country, still distinct and separate from the city, maintained its traditions and agricultural ecology largely unchanged.

The hills behind Kowloon were always a barrier between the peninsula and its hinterland, as the wild ramparts of Lion Rock, seen here, suggest.

In China, revolution, civil war, and later the Japanese invasion cast ever darker shadows. Hong Kong was the haven most South China refugees aspired to reach—and countless thousands did. The refugees who fled to Hong Kong in the late 1930s were mostly destitute, and their sheer number caused major environmental degradation.

The New Territories, covering almost 1,000 square kilometres and with over 200 islands, brought a new grandeur to the Hong Kong landscape. Their 1898 population was probably at least 50,000, and the 1911 census recorded over 69,000 people in the mainland New Territories' 600-odd villages.

The New Territories' Punti, Hakka, and Tanka clans were proud of their heritage and suspicious of governments. However, they were soon to develop remarkable cooperation with the British. The colonial officials, far from exploiting them as had most Qing mandarins, were clearly even-handed; and, despite their strange land ownership laws and ignorance of *fung shui*, they were intent on improving rural conditions. Thus, as China slipped towards revolution and chaos, and, as its peasants suffered and perished, the New Territories' clans prospered as never before.

The government produced increasingly detailed maps of Hong Kong. The first, a 'two inch to the mile' map, completed between 1899 and 1904, provided the basic information for the New Territories' administration. A larger scale map, completed in 1905 and revised in 1909, gave the first truly accurate view of the entire territory. The maps showed how much the rugged terrain had limited access: in 1909 Lantau had only village pathways, while on the mainland—besides some 'Pack Roads and Paths'—only a single road led northwards.

The construction of the Kowloon–Canton Railway (KCR) brought vastly improved access to the New Territories. The line, primarily intended to strengthen trading ties between Hong Kong and South China, was begun in 1906. The British section (to the border at Lo Wu) was completed in 1910, and the much longer Chinese section (from there to Guangzhou) was finished in 1911—the same year the Qing Dynasty was overthrown.

The New Territories' terrain—steep and rocky or flat and swampy—made building the railway difficult and expensive. The 35-kilometre British section cost almost 50,000 pounds sterling per kilometre, then a very high figure—and it needed five tunnels, forty-eight bridges, sixty-six culverts, and extended embankments. Almost one million cubic metres of rock and earth were moved to form the culverts and embankments, and the two-kilometre tunnel under Beacon Hill was the longest in China.

The British section was opened on 1 October 1910. The line was intended to ensure Hong Kong's position as the leading trans-shipment port for South China, as one journalist enthused:

> *The Kowloon–Canton Railway is the first tentacle, the first artery through which the red blood of trade will flow to and from this centre of British interest.... It opens the interior of China to the greatest emporium in the East.*

But the KCR did more than that. It also broke through the topographical barrier—the Kowloon hills—that had hidden the hinterland from people on Hong Kong Island. For the New Territories' villagers the railway was a revolutionary intrusion, bringing both *fung shui* troubles and better access to the lucrative urban markets.*

Hong Kong's expanding transport network, reservoirs, reforestation, and much more, all stemmed from Western science and British administration. The lesson was not lost on the Chinese revolutionaries who, familiar with Hong Kong, were striving to draw China into the modern world. Moreover, from Hong Kong's increasing wealth, as an observer wrote:

> *Branches trail out, like long vines, over the plains of Guangdong, into the valleys of Fujian and Guangxi and further afield all over China.... Many a warlord and provincial official owes a sense of security to his shares, treasure, or real estate in Hong Kong.*

The tragic typhoon of September 1906 had emphasized the urgent need for a large typhoon shelter to protect junks and sampans. A smaller shelter, built at Causeway Bay in 1883, was now inadequate. Indeed, following severe typhoons in 1901 and 1902, the Hon. Gershom Stewart, a member of the Legislative Council, had already urged that a new shelter be built at Yau Ma Tei. As Stewart said on 14 December 1903, without the protection of typhoon shelters, Tanka 'men, women and children have nothing between them and the next world but perhaps a half inch plank when it may be blowing a hurricane in the harbour'.

The government, while agreeing in principle, claimed that funds were not available for the work—an excuse Governor Sir Matthew Nathan (1904–7) repeated exactly a week before the typhoon struck on 18 September 1906. For the thousands of boat people who perished that morning the procrastination had continued too long. As a poem published soon after in the *South China Morning Post* put it:

> *Rain that strikes like the driven hail,*
> *Is it wind or spirit that tears the sea?*
> *A cry and the glint of a riving sail,*
> *My boat, my boat, and my babies three ...*

Two days after the typhoon Nathan told a sombre Legislative Council:

> *Hong Kong has just suffered a catastrophe as calamitous, if not more so, than any which has previously befallen the Colony.... What happened to the Chinese boats was evidenced by the appalling scenes of desolation along the prayas and the Kowloon shore.*

Yet—incredibly—a decision to build more shelters was still pending two summers later. The government was cowed by trading companies which resisted a paltry tax to finance the work, though it was the boat people who lightered their cargoes. Finally, after more deaths during a typhoon in 1908, the government acted: by setting a marginal tax on ships entering the harbour, it quickly raised sufficient funds.

Thus was the first Yau Ma Tei typhoon shelter born: a solid causeway that formed a reasonably safe enclosure for boats. Governor Sir Francis May (1912–19) laid a commemorative stone to mark its completion on 16 December 1915: to Hong Kong's shame, twelve years and two days had passed since Gershom Stewart had first raised the matter.

The First World War demanded major economic adjustments in Hong Kong but left the Colony physically unscathed. Despite local political and social tensions during the 1920s, the colony continued to prosper and grow. In 1911 the population was about 450,000; by 1925 it had reached about 725,000; and by 1931 it was almost 850,000. The number of ships using the port doubled about every fifteen years.

* *Since 1841 only the villages close to Victoria had prospered from selling food and fuel there. The value of the railway can be gauged from the fact that, when a branch line was opened to Sha Tau Kok in the north-east New Territories, the previously remote area soon prospered.*

By the 1920s Hong Kong Island, where it faced the central harbour, was largely urban, and more land had been reclaimed. Victoria was no longer the place that G. R. Sayer knew just after 1900, when its streets were:

> *the unchallenged hunting ground of the pedestrian, the chair coolie and the common carrier humping his load.... Traffic signals had yet to replace shady banyan trees, and the clatter and screech of gears and brakes the sob of the straining coolie.*

As late as 1900 Tsim Sha Tsui (Sharp Sandy Point) was still a sandy beach popular for launch picnics and bathing; but by the 1920s the shore had been reclaimed for the KCR terminus and wharves. The paddy fields and most of the hills that once spread across southern Kowloon had gone, replaced with businesses and residences. Nathan Road was an elegant banyan-lined boulevard, reinforcing the view that Kowloon was an unspoken European 'reserve'. But on both sides of the eastern harbour the foreshores were still little changed: Lei Yue Mun and Shau Kei Wan were small fishing villages surrounded by open hillsides.

By the 1920s Hong Kong's urban conditions were, if far from ideal, much improved. Over-crowding led to extreme squalor in localized areas, but the population was too small—and too frugal—to degrade the wider environment. Industry and motor vehicles had still barely affected the colony. Hence, despite drifting smoke from countless wood or charcoal cooking fires, and coal smoke from power stations and ships, the air was still largely unpolluted. Nor was there any general decline in the water quality, despite the filth around sewer outfalls and coastal villages.

With unpolluted skies the Peak gave stunning views. A road was built there in 1922, but the place still had drawbacks. The moisture remained abysmal: 'damp worse than Nigeria', complained Governor Sir Frederick Lugard (1907–12), vexed by 'envelopes all glued together [and] cigars like bits of sponge'. The Peak Reservation Ordinance of 1904, in force until 1930, limited residency to Europeans of whom the Governor approved. A popular contemporary rhyme about the Chinese Francolin, a partridge whose rasping call was common in spring, lampooned the snobbery of Peak society:

> *Be you outdoors or indoors you can't get away*
> *From this bird which incessantly, tirelessly brags.*
> *Who first was the wretch with the disarranged brain*
> *That instructed the bird to declaim the refrain*
> *'Come to the Peak. Ha! Ha!'*

Ships steaming into Hong Kong via Lei Yue Mun in the 1920s and 1930s passed islands little changed from a hundred years before: 'severe and rugged, treeless and covered with short grass', as the author of *In Far Eastern Waters* (a guide to Hong Kong published in the 1930s) saw them. Even the eastern harbour was unchanged, with its denuded and empty hills rising behind Kowloon Bay. But around Victoria the scene was fundamentally different:

> *The main city [stands] proudly on the lower slopes, with its big residences perched on the topmost peaks.... The hills above the town are clad with a thick carpeting of trees.*

The reforestation initiated by Governor Kennedy and Charles Ford half a century before had born fruit. The urban ecologist C. Y. Jim estimates that by the early 1900s a quarter of the Territory was covered with woodland—and, as he notes, 'no effort was spared to preserve the water catchments'. By the mid-1930s, as photographs show, Mid-levels and even the Peak were relatively well wooded; and foresters had planted countless ornamental trees in the developed areas. In early summer Flame of the Forest trees, natives of Madagascar, flowered brilliantly in scarlet and vermilion. The first Flame trees had been planted in 1908, and took some ten years to mature into flowering trees.

Hong Kong's post-1900 reforestation was inspired by two tireless botanists, who between 1903 and 1920 were successive superintendents of the (renamed) Botanical and Forestry Department: Stephen Troyte Dunn and William James Tutcher. They realized that the key factors controlling Hong Kong's vegetation were its extreme seasons, and the destruction wrought by summer typhoons and winter fires. Moreover, they concluded that Hong Kong had once been well wooded. Hence, but for the villagers' 'perhaps excusable custom' of fuel gathering and hill burning, in the long term Hong Kong's reforestation could become self-sustaining.

Birds gradually colonized the new woodlands around the Peak, as the Botanical and Forestry Department's 1907 report indicates. That year the department consulted J. C. Kershaw, an ornithologist, 'with a view to encouraging the increase in singing birds which have become so cheering a feature of the Botanical Gardens and surrounding portions of Hong Kong in recent years.'

The government's annual reports for the 1920s chronicle the continuing effort to 'green' Hong Kong. Upwards of 200,000 Chinese Pines alone were planted out in most years. Broad-leaved trees were also planted, and at Tai Wo (near Tai Po) over 6,000 Camphor Trees were established. Creepers that threatened to strangle young trees were cut back, and a war was waged on

The Shing Mun (Jubilee) Reservoir, completed in the 1930s, was then the largest water scheme in the British Empire. It's surrounding hills were then severely denuded and eroded.

caterpillars infesting Chinese Pines. New fire barriers were cut and, before the 1924 to 1925 dry season, old fire barriers were cleared of grass along some 70 kilometres. Elsewhere undergrowth was removed to destroy the habitats of anopheles mosquitoes.

The burnt offerings left after grave worship remained a major cause of hillside fires, especially during the Ching Ming and Chung Yeung ancestor festivals. Falling in the third and ninth months of the lunar calendar, these festivals came before and after the summer rains when the hills themselves are usually dry. Thus, in 1926, government forestry workers cut fire barriers around 'every grave on the hills above Shek O'. There were no fires in government reforestation areas in 1927, but when other hillside blazes occurred 'little or no assistance in dealing with the fires was given by village people'.

Perhaps the villagers concerned resented the foresters' work for *fung shui* reasons; perhaps they saw no reason to fight fires where they, as farmers, often *chose* to burn the vegetation. Perhaps, indeed, they had *lit* the fires. Whatever the truth, their non-cooperation underlined a critical fact: Hong Kong's foresters could achieve only so much regeneration if enough people ignored—or, worse still, hindered—their work.

After about 1920 Hong Kong could no longer have survived without the New Territories. The reason was simple: water. By 1918 every Hong Kong Island valley with the potential for a large reservoir had been dammed, and water engineers turned their gaze to beyond Kowloon.

During the late 1920s various small reservoirs near Kowloon were brought into commission. By the mid-1920s the huge Shing Mun Valley Water Scheme was being planned, and work on it began in 1928. By then the population was well over 700,000, almost 200,000 more than when Tai Tam Tuk Reservoir had been completed just a decade earlier.

The summer monsoon failed disastrously in 1929. The rainfall during May and June was a quarter of the average, and almost all the reservoirs dried up. Hong Kong Island, where most people lived, was the worst affected. Royal Air Force planes attempted a

Hong Kong's growing need for water could only be solved by building more reservoirs. Tai Tam Intermediate Reservoir, shown here, was completed between 1907 and 1908.

These stone steps once led to one of the eight Hakka villages submerged beneath the Shing Mun Reservoir.

rain-making experiment, but failed. Water tanks were built along the harbour. Junks and sampans ferried water (brought by train from the Shenzhen River) across from Kowloon; barges brought water from Macau; and ships brought more from Shanghai, Manila, and Singapore. But despite these supplies probably at least 200,000 people fled to China.

Seven summers passed before the Shing Mun (Jubilee) Reservoir was completed. Opened in 1936 or 1937, it was then the largest water scheme in the British Empire, and it flooded the deep Shing Mun valley set below Tai Mo Shan. Eight Hakka villages and some 70 hectares of fields were submerged, and 855 villagers were resettled elsewhere (though many later worked for the waterworks). Two dams enclosed the valley and held back some 13.5 million cubic metres of water, twice Tai Tam Tuk's capacity.

By then Hong Kong had thirteen reservoirs. This huge capital investment had a total capacity of 27 million cubic metres. Even so, in the late 1930s population pressure meant the water supply was often worrying.

About 1910, as G. R. Sayer wrote, 'Victoria still virtually comprehended everything on the Island'. Repulse Bay was 'remote and unravished', the village of Tai Tam Tuk was 'unconscious of its coming doom'—and the only way to Deep Water Bay and Shek O was by boat or on foot. Boat picnickers felt 'a twinge of disappointment if a second launch disputed possession of one's chosen beach'.

By the late 1920s that was much changed. Roads encircled Hong Kong Island. In the New Territories, besides the railway, a road led from Kowloon to Tuen Mun (Castle Peak), and from there on to Yuen Long, Sheung Shui, Fanling, Tai Po, Sha Tin—and so back to Kowloon. For walkers, reservoir catchments provided ever more extensive trails through otherwise often inaccessible country. Only the outlying islands remained difficult of access.

In compact, urban Victoria it perhaps seemed barely possible that the road encircling the New Territories stretched for over 90 kilometres, or that it meandered through agricultural lands whose Punti and Hakka villagers tilled the soil largely as their ancestors had. For, as *In Far Eastern Waters* noted: 'The New Territories are

Chinese Fan Palms were still being cultivated in the 1930s. Among other uses, their leaves made roofing, hats and rain capes.

a miniature South China, but much more prosperous, and more tranquil, than most of the interior [of China].'

In Far Eastern Waters described the road tour around the New Territories. West of Kowloon the grandeur of Tai Mo Shan and Lantau dominated the landscape ('like the wilder parts of Scotland'). But from Yuen Long to Sheung Shui a wide plain was patterned with paddy fields centred on Kam Tin, with its 'five ancient villages, defended by moats, and approached by quaint gates and stone bridges'. From Fanling ('tired businessmen find rest here from the rush of town life') the road led on to Tolo Harbour and Sha Tin. There mountains and water combined lyrically, and the local rice was some of China's best.

The past half-century had brought significant changes to the New Territories, especially since 1900. There had been a slight shift from rice growing to vegetables, as the latter were more profitable. The urban markets were now accessible, and the population there provided abundant manure. During the 1920s and 1930s many refugees became tenants to absentee land owners.

But much more remained unchanged: the rural calendar's cycles, the division of tasks between men and women, the harvesting of the uplands for fuel and herbs, the use of only natural materials. Above all, rice paddy still dominated the lowlands—and the land was patterned by traditional rural beliefs.

The Fanling scene witnessed by G. A. C. Herklots in the 1930s had changed very little in generations—even centuries. It was July, and the first rice crop was drying around the villages. Farmers, in preparation for the second crop, were plodding through liquid paddy fields behind buffalo. One beast was dragging a long-pronged harrow, with a boy sitting on it to drive the prongs deeper into the mud: 'He was gazing at the clods of earth in front of him, and holding on firmly lest he should slip forward and get mangled'.

Alexander Grantham and his wife Maurine often walked through the New Territories in the 1930s—'All day, up hill and down dale'. The mountains gave majestic views over the surrounding country, and the valleys were rich with agriculture. Grantham, fluent in Cantonese, came to appreciate the land and its people:

> *We passed through tiny villages where I would chat with the farmers. We followed ancient paths paved with heavy granite slabs, old as the hills themselves. We came on miniature temples far away from any habitation where travellers might rest or offer incense to one of the gods. At these we would leave some sticks of incense burning on the altar. Occasionally we espied a very small shrine in an old gnarled tree, indicating the abode of a tree-spirit.*

Thomas Dealy was another lover of the countryside. A master and later headmaster at the venerable Queen's College, Dealy captured the countryside—and the simplicity of its rural life—in poetry:

> *Behind the sailing clouds, the falling sun*
> *Sends streams of light that patch the seas with blaze*
> *Of shifting dull red gold. A purple haze*
> *Enwraps the many distant isles. The dun*
> *Hills, clad with fragrant pine, have now begun*
> *To catch the sunset glow; athwart the rays,*
> *The burdened coolie down the hill-side ways*
> *Hies slowly home—his daily toil is done.*
>
> *The white-eyed* wa-mis *call from rock and tree*
> *In clear sweet notes of echoed rivalry:*
> *Belated* pak-hoks *come in stretching flight*
> *Unto the darkening land, and, in the light*
> *That swiftly wanes across the drowsy sea,*
> *Their wings and bodies gleam a ghostly white.* *

Hong Kong is very rich in insect species, with 200-odd butterflies and 90-odd dragonflies and damselflies. Even today butterflies are easily the most common companions on countryside hikes, so it was not surprising that one of the earliest studies by a local naturalist concerned butterflies—or that the beauty and variety of local species inspired a work rich both in science and in superb engravings. J. C. Kershaw, who published his *Butterflies of Hong Kong* in 1907, was clearly a dedicated perfectionist.

By early this century Western botanists knew that China had the world's richest temperate flora: some 15,000 species, half of them found only there. Hong Kong itself boasts almost 2,000 native species.

The leading Hong Kong botanists early this century were Stephen Dunn and William Tutcher. They explored for plants in Hong Kong and Guangdong, and in 1912 published *Flora of Kwangtung and Hong Kong*—a 400-page work which updated Bentham's *Flora Hongkongensis*. As the new work's title implied, it saw the Hong Kong flora as part of the wider Chinese story.

* *Dealy's poem appeared in the first issue of* The Yellow Dragon, *the Queen's College magazine. According to Gwenneth Stokes, my mother, it was 'the finest poem in English' ever to appear in that illustrious magazine.*

During the 1930s naturalists explored the uplands, such as here, near Ma On Shan.

Lantau Island, seen from Lamma during late afternoon. The combination of light and water in Hong Kong is often entrancing—as Thomas Dealy's poem suggests.

Numerous other botanists drew on Dunn's and Tutcher's work. One was A. H. Crook, the most learned of Queen's College's headmasters and a noted authority on local birds. In 1930 he produced *The Flowering Plants of Hong Kong*, a small but painstaking book with the first published illustrations of local flora. A scholar at heart, Crook nevertheless gave the few common names then existing for Hong Kong plants. But, as he noted wryly:

> *There are really no 'common names' for most of the plants here. When a person takes up even a very common plant you have to say, 'That is* Stachytarpheta indica*', and the person simply gasps or sneezes or says something quite unprintable.*

During the 1930s Hong Kong's naturalists were almost all gifted amateurs. Most had specific interests but a few, notably G. A. C. Herklots, studied the entire lowland and upland ecology. The botanists' and naturalists' discoveries showed clearly that Hong Kong was rich in fauna species. However, the populations were often small and localized, thus making them highly vulnerable to human interference.

Centuries of agricultural deforestation had killed or driven off the large mammals once native locally—and those that occasionally wandered in from Guangdong, usually during winter, were often killed. Hardly surprisingly, sightings of South China Tigers and Leopards became steadily rarer; and by the 1930s few people could believe that such magnificent creatures had ever been native locally. One of the last tigers seen in Hong Kong roamed around Tsuen Wan (then an agricultural town) during the winter of 1934 to 1935. As Herklots writes, on one occasion,

> *An old woman grass-cutter was returning home from Tsuen Wan when the tiger walked up to her and started to circle her. She was terrified, but when it came too close she summoned up courage and gave it a few blows with her pole and managed to scare it away. When interviewed later the woman was still in hysterics.*

Especially in remote villages, encounters with Wild Boar and venomous snakes were much more common. Mammals which damaged crops, including Civets, Barking Deer, and Porcupines, were regarded as vermin—and killed. Chinese beliefs that some animals have medicinal properties remained potent and destructive, and were condemned by most Europeans. The Chinese, for their part, wondered why some Westerners found recreation in killing animals. There was a fox-hunt based at Fanling; and shooting was often enjoyed, despite the local difficulties that one young bank clerk recorded:

> *Towards noon, when the day becomes stifling, snipe shooting is not easy, for the heat seems to rise from the earth in quivering steamy columns.... The sportsman has to walk [on the paddy bunds] and is continually slipping into the hot mud.*

Since 1913 the government had gradually extended its plant preservation ordinances, and by 1936 almost a quarter of the local flora—and almost all the rare species—were 'protected'. A half-hearted ordinance protecting two mammals (pangolins and otters) was introduced in 1936, and there was a limited bird preservation ordinance. But between official dictates and rural traditions there was a gulf of values—and of actual enforcement.

The conservation reserves that were becoming popular elsewhere could never have been established in the Hong Kong of the 1930s. Villagers still occupied or used most of the countryside, and the urban people were mostly desperately poor. Yet the government could easily have initiated serious research into local ecology and agriculture. However, excluding the efforts of amateur naturalists and government botanists, virtually nothing was done.* Hong Kong—compromised by influxes of refugees and by the government's meagre attempts to foster a sense of belonging—was becoming an increasingly rootless place.

1937 was a black year for Hong Kong. The population had grown by some 150,000 in five years and in 1937 it probably passed one million. Housing had never been able to keep pace, partly because of the lack of land. By 1937 urban overcrowding was critical: three to four times the usual number of people lived in some tenements, and countless shacks had been thrown up on hillsides. The water supply and sewerage posed daunting problems.

Government precautions and immunizations minimized the risk of disease, but Hong Kong could not be quarantined. Refugees were flooding in from China where disease was rife—and the germ-laden Pearl River swept into Hong Kong waters. In the summer of 1937 the inevitable happened: cholera reached Hong Kong from Guangdong.

Sickness spread rapidly through the poorest areas. Medical staff and volunteers worked tirelessly to allay panic and control the epidemic. One sailor wrote home on 14 August: 'Emergency inoculations have been instituted, [and] the Roman Catholic officials have issued a decree allowing Catholics to eat meat on Fridays instead of fish, because the latter is the chief source of infection.' Over 1,000 people died before the cholera was contained.

As if to purge Hong Kong, a severe typhoon hit soon after the cholera outbreak. The fast moving storm erupted before dawn on 2 September, with winds that exceeded 200 kilometres per hour and the lowest barometric pressure ever recorded in Hong Kong.

The physical destruction was immense. As the *South China Morning Post* reported:

> *Huge waves surged over the Praya, reaching almost to Queen's Road.... A score of [ships] broke loose from their doubled moorings and careered drunkenly about the harbour in macabre dance.*

About thirty of the 100-odd ships in port went aground, and of some 3,500 junks and sampans almost 2,000 were sunk or wrecked. More would have been lost but for the Yau Ma Tei shelter. The loss of life was tragic: estimates ranged up to 11,000, most of them either boat people or refugees living in shacks.

Meanwhile, the Japanese had invaded Manchuria in 1931. Their forces marched southwards from Beijing during 1937, and in 1938 they captured Guangzhou. For China the Japanese invasion and the concurrent civil war were dire—and their effects in tiny Hong Kong were catastrophic. T. J. Ryan, a Jesuit serving here then, wrote: 'Neither the scourge of sickness or storm was as terrible in its ultimate consequence as the third that [1937] brought—war.'

Hong Kong faced a dramatic decline in its entrepôt trade—and a relentless flood of humanity. About 100,000 refugees arrived in 1937, 500,000 in 1938, and 150,000 in 1939. By 1941 the population was over 1,600,000. The government and community responded with compassion and generosity for these destitute, starving people—and with faith in Hong Kong's long-term future.

The refugees wrought environmental havoc. By 1937 the reforestation was more extensive than ever before, despite rural

* *Many of these discoveries were documented in* The Hong Kong Naturalist, *a journal published during the 1930s. G. A. C. Herklots and G. S. P. Heywood made detailed, invaluable descriptions of the Hong Kong countryside during the 1930s. However, because of the Second World War, Herklots' and Heywood's books were not published until 1951.*

villagers still exercising their customary right to gather fuel, and some illegal tree-cutting by the urban population. But after 1937 Hong Kong looked on helplessly as refugees, and those exploiting them, scoured the hillsides for fuel. In January 1938 an 'Old Resident' wrote with feeling to the *Hong Kong Telegraph*:

> *Thousands of pine trees have been cut down in the New Territories.... Vast areas, where but a few years ago trees were flourishing, now present a pitiful vista of stumps. If this ravaging is to be stopped, and surely it can be, severe penalties must be imposed upon those who are engaged in the trade. Furthermore, the number of forestry guards must be increased.... To come across a tree-cutter a few years ago was unusual, and a warning shout was enough to send him scurrying away. Today, these people are to be seen working in small gangs, and if spoken to show a brazen disregard of any protest.*

Hong Kong's rural people at least had harvested the hills with some thought for their future needs—and for the land itself. But the refugees had little regard for the future or their surroundings: to them Hong Kong meant only immediate survival.

By the late 1930s a line of fortifications stretched across the central New Territories. The defences, especially those around the Shing Mun Reservoir, were major works and impacted on the countryside. On the Stanley peninsula a road had been pushed up into steep boulder-strewn slopes, and a fort built on a flattened hilltop. Completed in 1937, Stanley Fort led a Royal Engineer to comment: 'One of the most amazing things to a newcomer to Hong Kong is the apparently airy manner in which an engineer will say, "We'll just take the top off that hill and fill in the valley here."' In the decades to come wonderment was to merge with awe as Hong Kong recast its landscape.

The colony's centenary in 1941 was muted, as by then Japanese troops were camped along the Shenzhen River. However, some of Hong Kong's grand old men gave a series of radio talks.

Sir Robert Ho Tung, then seventy-nine, recalled the time when Kowloon was still part of the Chinese Empire and covered with paddy fields. Sir Shou-son Chow, aged eighty-one, whose local ancestry stretched back seven generations, noted that there were now about 6,000 motor vehicles in Hong Kong, and good access to the surrounding countryside. He stressed the community's impressive achievements, 'a hard-won heritage which we should treasure'. And Sir Robert Kotewall stated:

> *If some Rip Van Winkle of the Island, fallen asleep a century ago, were to emerge from a long hibernation he would be unable to recognize the place as he last saw it. Then, it was a barren island, with scant vegetation and a tiny population. Now, it is a city of imposing structures; of great shipyards and wharves; of wide roads along which traffic ceaselessly pours; of teeming concourses ... set amid wooded hillsides and fronting a harbour on whose waters lie ships of the seven seas.*

The transformation, Sir Robert believed, was the result of cooperation between 'the British, with their enterprise, foresight, and initiative, and the Chinese, with their capacity for hard work, patience, and adaptability'.

Hong Kong had come through the First World War unscathed by the devastation of conflict. A generation later such was not to be its fate. And, as the summer of 1941 turned to winter, people prepared for the expected invasion. When war finally engulfed Hong Kong on 8 December 1941 its people suffered dearly—and the landscape was ravaged and laid to waste.

Even shrubs, such as this Melastoma candidum, *were cut for firewood by the refugees who swarmed into Hong Kong in the late 1930s.*

Beyond Tuk Ngu Shan and High Island Reservoir, the sun sets behind Ma On Shan's saddle-back summit.

REGION

SAI KUNG PENINSULA

It was as though the earth, intolerant of restraint, had reared itself up in protest. As a delightful corollary, where these broken hills plunged into the sea or stepped delicately down to the shore, they carved a coastline infinitely sinuous and brimful of variety and charm.

G. R. SAYER, 1975

It is midsummer. Golden light bathes the summit of Tuk Ngu Shan, a hill above the eastern end of High Island Reservoir. To the west the sun is slipping down past Ma On Shan's saddle, its last rays casting a sheen over the reservoir. The wind has softened to a breeze.

Planning only a quick, hilltop reconnaissance, I have left my overnight gear a few kilometres downhill. Now I am too far away to fetch it before dark—but nor do I wish to tear myself away. The afternoon clouds have gone, and a sky of deepening indigo promises a brilliant dawn. So I stay: I have my camera gear, a few oranges, and some water.

In the gathering darkness Sharp Peak's conical summit juts up above the ridges to the north. Serried ranges stretch to the south towards Kowloon and Hong Kong Island, some twenty kilometres away. The beacons of Victoria Peak are red pinpoints; and the topmost lights of Central Plaza, Hong Kong's tallest building, are just visible.

Later, sheltered among shrubs, I bed down on some hastily gathered grasses. There is total silence. The lights of aircraft, so distant that they drift noiselessly through the sky, are the only reminder of urban Hong Kong. I sleep fitfully. Sometime before dawn, a single aircraft flies directly overhead. Amid the silent hills, the noise is so startling that I wake in fright.

Daybreak comes at last. In the pre-dawn light cumulus clouds tower above the horizon. Then the sun, a brilliant disc, rises above Mirs Bay. Rays shaft down, spread, and turn the sea to gold.

G. R. Sayer, quoted above, was describing the symbolic dawn of 1898: when Hong Kong's hinterland, since time immemorial part of China, was at a stroke attached to the Colony. Its territory now included not only Hong Kong Island, Kowloon, and Stonecutters Island, but also far grander, wilder country—the New Territories, 'brimful of variety and charm'.*

Though much of the New Territories was remote and rugged, its lowlands were far from empty. Tracks criss-crossed the land, linking village to village, valley to valley. However, early this century there were no roads. Indeed even in the 1930s the eastern New Territories—the wide peninsula that stretches north-east past Sai Kung—had not a single road.

This north-eastern region beckoned powerfully to naturalists and walkers during the 1930s. One of Hong Kong's loveliest areas, it had rugged mountains, deep valleys, and indented coasts. And, except around the scattered villages, it was wild and challenging.

The area is still mostly empty and unpeopled. Three Country Parks together cover over 10,000 hectares of country: Ma On Shan Country Park, Sai Kung West Country Park, and Sai Kung East Country Park. The MacLehose Trail, which begins near High Island Reservoir and winds on to Tate's Cairn, runs for almost 40 kilometres through the three Country Parks—yet crosses just two public roads.

Tate's Cairn is one of the peaks behind Kai Tak Airport. From there tracks lead northwards over Buffalo Hill, and down to the uplands between Sha Tin and Sai Kung. East of the Ma On Shan massif, the land falls steeply to the narrow neck of the Sai Kung peninsula. Beyond there lower hills run out to High Island Reservoir in the south-east, and to Sharp Peak in the north-east.

* *The G. R. Sayer quotation comes from his second volume of Hong Kong history, covering the years 1862 to 1919. The work was published posthumously in 1975.*

The hills above Sai Kung are steep and rough. Scalloped volcanic rocks frame Buffalo Hill, seen in the background, where old Hakka terraces climb the slopes. Beyond Buffalo Hill's saddle, the tip of Kowloon Peak is just visible.

From Tate's Cairn (Tai Lo Shan) one can look down into the concrete jungle of industrial Kowloon—or along the uplands of Ma On Shan Country Park. A high chain of peaks runs to the north-east: to Buffalo Hill (Shui Ngau Shan), 606 metres; to Tai Kam Chung (Pyramid Hill), 536 metres; to Ma On Shan, 702 metres; and so on to The Hunch Backs (Ngau Ngak Shan). South of Pyramid Hill there is an undulating plateau. Everywhere else, except around the passes, the slopes are sheer, the ravines precipitous, and the ridges razor-sharp. This volcanic country is not for the careless or the weak.

Hakka clans settled these uplands. Their name, 'guest people', belies their enduring presence: throughout the uplands of the New Territories their pathways still traverse the hills. Some are earth tracks, mere byways between remote villages; some are laid with boulders, substantial lasting routes. The time and toil to make them were prodigious. Was it because their hills were so rugged, or in spite of them, that the Hakka became renowned for their physical stamina and their doggedness?

The 1950s scene Han Suyin captured in *A Many Splendoured Thing*, the terracing of slopes for modern buildings on Hong Kong Island, might well have been set some three hundred years ago, when Hakka clans first 'opened' the New Territories' uplands and carved out their fields:

> *Scores of black-clad Hakka women, wearing their crownless straw brims fringed with black cloth, carry off, in ant-like procession, basket after basket of soil. The top of each hill is slowly eaten off, and the earth appears, a bright ochre patch amid the surrounding green bush.... Finally the earth is stamped flat with their bare feet and wooden hand paddles.*

Hiking through Hong Kong's eastern uplands today, the land seems workable for grazing but far too steep for cropping. Yet on almost every hillside, half-hidden under tall grasses, are the remains of stone-walled terraces. Tea and indigo grew on higher terraces, peanuts and vegetables lower down, and rice wherever water was plentiful.

Did tea bushes really once spread across these now-grassy hills? 'Tea is cultivated in several places and is generally called *shan cha* (mountain tea)', Rudolf Krone, a German missionary, reported in 1858. 'It has rather a strong astringent taste, but is much liked by the natives, and particularly by those of advanced age, who consider that it promotes digestion and cools the system.'

Until well into this century the Hakka were still cultivating hemp: *wong ma*, or yellow hemp for rope, and *ch'ue ma*, the hemp for spinning. After harvesting, the hemp was dried, then its outer fibres were spun on primitive wheels. Itinerant Hakka weavers, most of them men, wove the material; village women later dyed it with indigo and made hemp clothing and bedding. All this endured until only about two or three generations ago—near where textile workers now make materials on computer-controlled looms.

The Hakka uplands were among the last parts of rural Hong Kong to be settled, and it was the Hakka who probably deforested Hong Kong's least accessible native woodlands. On the steeper slopes, where subsequently the old soils eroded badly, grassland and shrubland still dominate the scene. But in the high sheltered valleys, which lost less of their soil after being deforested, woodlands have grown back. After the Hakka villages were depopulated and abandoned in recent decades, woods have spread naturally from the villages' *fung shui* groves.

'The hills still lay like ancient dragons coiled in an eternal sleep', a visitor wrote around 1900, describing the line of peaks behind Kowloon. A rugged, precipitous barrier, the range shut out the hinterland. The New Territories were unseen, mysterious—except to people who climbed over the passes. Not surprisingly few Europeans ever ventured across the hills.

The Kowloon–Canton Railway (KCR) broke down this centuries-old barrier in 1910. The 35-kilometre line to Lo Wu demanded major earth works throughout its length—as the country was either too high or too low, too steep or too swampy.

Beacon Hill, west of Tate's Cairn, posed the KCR's greatest topographical challenge, as R. J. Phillips has documented. No train could climb over its pass, so a two-kilometre tunnel was cut through the hill. Near the entrances, the ground was so soft that it was prone to collapse, but too hard for pick-and-shovel excavation. Further in the rocks were often deeply weathered and rotten. Streams and springs permeated the hill, inundating the tunnel and making the weathered granite treacherous. But elsewhere the rock was so hard that up to 40 kilograms of dynamite were needed to blast through a single metre.

Across the agricultural lowlands the line—like some mythical creature—snaked along high embankments that wound above a patchwork of villages and paddy fields.

The second paddy crop was ripening around the 600-odd New Territories villages when the KCR opened on 1 October 1910. Pulled by an engine puffing coal smoke into a hazy autumn sky, an inaugural train took dignitaries to Lo Wu and back. Most of the guests had never been beyond Kowloon. Emerging from the Beacon Hill Tunnel they were suddenly transported to a much older landscape—of majestic peaks, tranquil inlets, clustered villages, and *fung shui* groves. A *South China Morning Post* reporter described the country near Sha Tin:

> *The hills reflect their shadows in placid waters. Villages nestle in the shade of clumps of trees, and healthy-looking paddy fields give promise of good crops. Food for the sportsmen's aim takes wing as the train moves along, cattle are browsing and buffaloes are wallowing, while the duck farmer minds his flock that they may not cross the track of the iron horse.*

Some ten years later the scene was unchanged in all its essentials. 'Once through the tunnel you are back in biblical times', one woman traveller wrote. She continued:

> *The Sha Tin valley was being harvested by women wielding sickles, while a small portion of the earlier ripening rice was being threshed by hand. A few buffalo and cows tended by children grazed at the foot of the railway embankments, while the roads where they paralleled the line were deserted but for the occasional drover with half a dozen animals or a swarm of ducks. On the water the odd sampan propelled by the* yu lo *[stern] oar could be seen, and in distant Tolo Harbour a junk or two was under sail.*

The KCR thus revealed the New Territories' rich lowlands. Although more distantly, it also revealed the rugged uplands. R. C. Hurley, who had been roaming Hong Kong since 1879, left this description:

> *The beautiful Sha Tin valley [is] hemmed in by high land on every side, with Tide Cove and Tolo Harbour reminding one of the lochs of Scotland.... Sha Tin Station is quickly cleared, and for seven miles the line skirts the shore opening up ever-changing and beautiful vistas at each sudden bend.... The Hunch Backs and the twin peaks of bold Ma On Shan are seen across the Cove.*

The Hunch Backs and Ma On Shan still dominate the Sha Tin valley. They are rugged, wild, and hazardous. A knife-edge track

Autumn grasses on the uplands above Sai Kung.
Pyramid Hill and a mostly hidden Ma On Shan form the backdrop,
close to where Heywood wrote of 'mountain tops and wide sky'.

leads over The Hunch Backs, past slopes too sheer for comfort. When I was last there, a friend and I were delayed by mist, hiking homewards. Dusk faded to darkness and, brought up by a dangerously cliffy drop, we bivouacked on the track. It seemed uncanny. We were surrounded by the same rough country the Hakka—and Hurley—knew, but below us were the massed lights of Sha Tin New Town.

The next morning we looked over to where we had been the day before: on Ma On Shan's buttress-like spurs and its 702-metre summit. The saddle between the two highest points, for which Ma On Shan (Horse Saddle Mountain) is named, stood out. Below there, on its north-eastern face, are ravines which even the Hakka tree-cutters probably never harvested. There rich woodland grows today, with ashes, oaks, laurels, camellias, and rhododendrons.

Little wonder, I thought, that the country near Ma On Shan beguiled the naturalists and walkers who explored Hong Kong during the 1930s—when Barking Deer were still often seen through these hills.

'Vinjar'—pen name of V. H. C. Jarrett—wrote a newspaper nature column during the 1930s. On an April day with persistent rain 'Vinjar' and G. A. C. Herklots, the naturalist, went looking for spring-flowering rhododendrons on Ma On Shan. 'Hacking and clambering, we eventually reached a grass-cutter's path and then commenced a steeper ascent', he wrote of the thickly vegetated ravines. It was late afternoon when they reached the summit, pleased to find rhododendrons in full bloom. There were prolific Red Azaleas and some mauve Farrer's Rhododendrons. 'The prize of the climb' were some rare white-to-pink Champion's Rhododendrons—first described by John George Champion after another spring excursion, on 18 April 1849.

Graham Heywood was perhaps Hong Kong's best-known climber in the 1930s, and later wrote a lively book describing his experiences—*Rambles in Hong Kong*. Ma On Shan, Heywood said, 'is a fine climb by any route, and on a clear day the view from the summit is unsurpassed'. Getting there required taking the train to Sha Tin, crossing Tide Cove (now a polluted concrete channel) in a sampan, and squelching through a mangrove swamp (now a housing estate). Then came The Hunch Backs, 'a most airy and exhilarating climb ... [where] nailed shoes give a sense of security lacking in the handholds of straggling grass'.

One wonders what the Hakka thought of these Westerners who apparently delighted in climbing for no obvious reason, even in bleak, wintry weather. Once, somewhere near Ma On Shan, Heywood was leading a party cross-country when clouds closed in:

> *We passed some cows placidly browsing in the fog, and continued to plod ahead, wishing we had a compass. It soon became obvious to the rearguard that the leader had lost the way. In a little while some more cows loomed up in the mist; they looked strangely familiar, and closer examination revealed that without doubt they were the very cows which we had passed a quarter of an hour before.*

Today, at least on clear days, buildings are always in sight below Ma On Shan and The Hunch Backs. But just south of Ma On Shan, near Tai Kam Chung (Pyramid Hill), is a wide, open plateau—and there one is still totally alone with nature. In Heywood's time a Hakka village stood nearby: 'A grove shades it in summer, and shelters it from the east wind in winter.' The village has long since been abandoned, and grown over with shrubs and trees. But the sense of boundless space and grandeur is unchanged. As Heywood put it, there one experiences 'only the mountain tops and the wide sky'.

Below Ma On Shan a deeply indented peninsula stretches eastwards: a landscape not of mountains and sky but of cliffs and sea. The wildest part of Hong Kong's mainland coast, the peninsula was for long kept remote by its rugged topography and exposed coastline.

Until 1945 the only access to Sai Kung was by a ten-kilometre track, and even in 1960 the region was largely cut off from urban Hong Kong. During the 1970s High Island Reservoir's construction greatly affected the area: the peninsula's remote villages were severely depopulated and many abandoned—and Sai Kung itself became a commuter town. Indeed, when Outward Bound established a base near Sai Kung in the early 1970s the area was, according to one Outward Bound leader, 'a remote and isolated place, ideal as a place to get young [city] people away from it all'. This soon changed; and when Sai Kung developed into a commuter area, Outward Bound resolved to build a sailing ship—to find adventure and challenge on the seas!

Even so, Sai Kung West Country Park and Sai Kung East Country Park cover almost the entire peninsula beyond Wong Chuk Wan—and beyond Pak Tam Chung the country is still virtually uninhabited. What Heywood wrote of the peninsula in the 1930s is still largely true today:

> *Sea and land meet in a tangle of headlands, inlets and islands.... Here and there is a fishing village in a sheltered bay, the houses crowding close together for company.... Apart from these you may go for miles without seeing anything of man's handiwork, for the hillsides fall steeply to the sea and there is little room for cultivation.*

The eastern Sai Kung peninsula and its islands form Hong Kong's most intricate coastline, seen here from Ma On Shan.

Shrub-covered hills, abrupt headlands, and beguiling islands typify the south-east of Sai Kung East Country Park—where Melastoma candidum *frames this scene.*

Sai Kung East Country Park's beaches are among Hong Kong's most beautiful. The dramatic peaks and wide sands seem almost Hawaiian here at Sai Wan.

This engine, the remains of a boat lost off Sai Wan, symbolizes the general rubbishing of Hong Kong's coasts.

The outer Sai Kung peninsula is a classic ria (or submerged) coast. The old coastal lowlands were long ago lost under the sea, and the ocean has since eaten away at the remaining land. The peninsula is almost unbelievably tortuous, with coves and bays, bluffs and headlands, islets and islands. The distance across its neck at Wong Chuk Wan is barely two kilometres—but the distance between the neck's two sides, measured around the coast, is over 90 kilometres.

The Sai Kung peninsula's hills are lower and more rounded than elsewhere in Hong Kong. There is nothing over 500 metres, though some of the higher peaks are still impressive—especially Sharp Peak which commands the peninsula's north-eastern extremity.

The vegetation changes from grassland and shrubland in the east to areas of woodland in the west, where it is wetter and more sheltered. The post-war government reforestation across the western peninsula was keenly supported by the remaining villagers. After the villages were abandoned in recent decades, the western peninsula's numerous *fung shui* woods propagated naturally, so enriching the plantation woodland with native species. Today native broad-leaved trees are maturing among now-elderly plantation pines, and steadily replacing them.

It is summer, and I am hiking around High Island Reservoir's southern side. The sun is scorching but clouds are fortunately spreading. With virtually no climbing, by noon I have almost covered the eight kilometres to the reservoir's eastern end. Around me shrubland slopes frame seascapes that are grand and wild: sheer headlands plunge into the sea, dark fingers of rock enclose coves and reach across inshore channels.

All around is rhyolite, a fine-grained volcanic rock which forms when lava cools rapidly. The rock's six-sided crystal structure, on a larger scale, creates hexagonal columns. The peninsula's eastern and south-eastern coast is dominated by rhyolite, with dark grey columns rising jagged out of the ground.

Later that afternoon, photographing from High Island Reservoir's sea-level dam, to my left are rhyolite headlands, their

Columnar volcanic rocks dominate the east coast of the Sai Kung peninsula, including Po Pin Chau, this 50-metre islet near High Island.

sides bare and angular. To my right is a rhyolite islet, its serrated sides only metres from the mainland. In front, with no land in sight, deep blue sea stretches to the horizon. Behind, the reservoir's main dam towers over the coast.

High Island Reservoir challenged the conventional rules of dam design. In the 1960s, with Plover Cove Reservoir, Hong Kong had already pioneered the concept of turning a quiet sea inlet into a fresh-water 'lake'. Now, a reservoir was to be created by building dams joining an island to the mainland—along a coast swept by ocean swells and typhoons. Such a recasting of the coastline demanded engineering on an almost god-like scale. Yet, driven by Hong Kong's acute need for more water, High Island Reservoir was completed in 1979.

Coffer dams were first built at sea level, the enclosed sea water drained, and the remaining salts flushed out. Then, after digging 25 metres down through the seabed to reach solid bed-rock, the main dams were begun. These two massive walls rise above their foundations more than 100 metres. When full, the reservoir covers 690 hectares and holds almost 273 million cubic metres—over 40 times the capacity of Tai Tam Tuk Reservoir.

Four hundred villagers, almost all of them Hakka, were resettled from the valleys around High Island. Five workers were killed during the work, as a memorial on the main dam records. But these losses brought lasting benefits. High Island Reservoir vastly increased Hong Kong's total water storage capacity—and formed the lake-like expanse that I saw from my unexpected overnight camp on Tuk Ngu Shan.

A few kilometres north of the reservoir, past hills that plunge down to inaccessible coves, is Tai Long Wan (Big Wave Bay). Other Hong Kong bays are perhaps more idyllic, perhaps more memorable. But nowhere else is there a sweep of water as majestic as Tai Long Wan, or vast ocean beaches so utterly removed from the concrete world of urban Hong Kong.

Seen on a map, Tai Long Wan's two arms look like the head and forelegs of a Tang horse—its rider's top-knotted head the promontory north of Sharp Peak, its tail the headlands facing Tolo Channel.

Seen from the hills to the south, Tai Long Wan extends blue-green across three kilometres of water. Its northern side is dominated by Sharp Peak: precipitous spurs lead up to the peak, and step down to an ocean headland. Around the bay tree-filled gullies decend to a fringe of sand and trees. Promontories reach into the bay, dividing it into three beaches. On calm days, the sea and sky merge along a blue horizon beyond Mirs Bay.

The islands lying off the Sai Kung peninsula are swept by largely unpolluted ocean waters—as this translucent rock-hole on Basalt Island suggests.

I have camped three times above Tai Long Wan, once alone, twice with friends. Each time the same, inescapable thought has recurred: how could anyone consider 'developing' so wild and beautiful a place? But that is the case.

When the Country Parks were established some small areas, mostly around old villages, were excluded from them—and one such area is near Tai Long Village, just in from Tai Long Wan. Recently, a developer acquired 50 hectares of this excluded land, planning to build a luxury resort there. It was also rumoured that the same developer induced the villagers to agitate for an access road through the Country Park. The developer's plans, challenged by environmental groups, remained the subject of various planning reviews during 1994. Then, in April 1995, the developer secretly landed bulldozing equipment at Tai Long Wan—and began excavating. The police intervened.

Undoubtedly some Hong Kong developers have scant regard for the public good—and less still for conservation. Stunned by this illegal excavation, Dr Ng Cho-nam, Chairman of the Conservancy Association, expressed a view that the government should urgently consider: that, without delay, it should compulsorily resume all such sensitive pockets of land surrounded by Country Parks; compensate the landowners; and include the land in the nearby Country Parks. Without such a bold initiative these irreplaceable wilderness areas may well slip through the planning appeal process—and be lost forever to 'development'.

From across Mirs Bay, a rainstorm advances towards Sharp Peak.

A dawn panorama from near Sharp Peak. Beyond Mirs Bay lies a Guangdong island 30 kilometres away.

For now at least, there always seems to be more of 'wild' Hong Kong to see. When I began this book the peak I most wanted to climb was Nam She Tsim—better known as Sharp Peak (named, not after its sharp peak but after Grenville Sharp, a financier). Its sheer slopes hint at an exhilarating climb; its twisted, conical summit, 468 metres high, promises remarkable views.

Go, if you can, to its northern side. Hike around from Tai Long Wan, past Sharp Peak and on to Ko Lau Wan, which is near Tap Mun Chau (Grass Island). If that is too far, take the *kai to* (local boat) from Wong Shek Pier, on the northern peninsula, to Ko Lau Wan. Spend a night on the headland there, looking over the sea to Sharp Peak's northern face and its rocky coves. The place is 30 kilometres from Wan Chai—but light years from urban Hong Kong.

Threatening sunset clouds symbolize the challenges of the post-war period.

HISTORY 1945–1970

A PHOENIX RISES

Refugees moved into the hills and hung their shacks in deep festoons over rocks bared by the wartime search for fuel. Always they crowded in on the town, for there alone lay the hope of rice for tomorrow.

'A PROBLEM OF PEOPLE', GOVERNMENT REPORT, 1956

The future is dark with planners' jargon, and much is heard about 'satellites' and 'urban over-spill'.... Driving through the New Territories you have the sense of making a journey through a doomed landscape. With modern earth-moving equipment valleys can be made straight and hills laid low almost overnight.

F. D. OMMANNEY, 1962

Hong Kong, crowded as never before with destitute refugees, faced critical challenges in the 1950s and 1960s. Nature added to the trials: for, in the years when the post-war refugee crisis posed its greatest threats, the summer monsoons often brought only a fraction of their usual rain—while in some years they swamped the land under awesome deluges.

In 1954 only half the normal amount of rain fell. The water supply was severely restricted, and the squatter shacks that sprawled over the hills became lethal tinder-boxes. Then, after two more very dry summers, incessant downpours flooded Hong Kong in May 1957. Storm-water drains burst, and tonnes of silt swept off the hills into the streets.

Nature was harsher still in the early 1960s. Devastating typhoons hit in 1960 and 1962. Then, after October 1962, Hong Kong entered its worst recorded drought—a two year ordeal that threatened the community's actual survival. The summer rainfall in 1963 was negligible. The May figure was 2 per cent of the average, and at the month's end Buddhist monks and nuns held a day of prayer for rain.

In June, at a time when the reservoirs would normally be overflowing with summer rains, they were only about one tenth full. The supply was cut to four hours once every four days; had water not been brought from China the rationing would have been much worse. The urban poor queued for water at standpipes; the farmers, whose crops depended on summer rain, faced ruin. 'Under the pressure of water-seeking everyone tastes the bitterness of an abnormal life', an observer wrote. As for animals and fish whose habitats were streams or paddy fields, they died in great numbers.

The Japanese invasion of Hong Kong had emphasized an evident fact of geography: Hong Kong was inseparable from the Chinese land mass. Ironically, the victory of the Chinese Communists in 1949 threw up a seemingly impenetrable 'bamboo curtain' along the Shenzhen River—a barrier that was as much an ideological chasm as a geopolitical border.

But the 'bamboo curtain' was far from impenetrable. Indeed, the émigrés and refugees who fled from China into Hong Kong during the late 1940s, 1950s, and early 1960s were the central fact of the Colony's existence. Between 1945 and 1970 they created unprecedented challenges and severe environmental degradation—yet it was their industry that transformed the Territory.

Hong Kong's traditional lifeblood, entrepôt trade with China, collapsed abruptly with the 1950 to 1951 trade embargoes that stemmed from the Korean War. Undaunted, the Colony restructured its economy in two decades to become a manufacturing exporter.

Industrial pollution began to affect the Territory; roads and reservoirs, housing and factories, spread into the lowlands of the New Territories. Meanwhile, tempted by urban wages, increasing numbers of villagers deserted their fields. As late as the mid-1960s, throughout rural Hong Kong, rice was still being grown in paddies ploughed by buffalo: a decade later rice growing had almost disappeared.

During the 1960s the government's annual reports all had covers with photographs showing aspects of Hong Kong's modern industry. The 1969 cover was an exception: it showed a New

A half-dead banana frond symbolizes the degradation of the natural landscape that war had brought.

Territories fish farm, 'typical of the quiet beauty of Hong Kong's hinterland'. It was the government's final annual report to open with an agricultural scene.

The Japanese occupation had devastated Hong Kong. Destruction from actual fighting was significant. Worse still, however, was the havoc wrought by almost four years of exploitation and neglect: services had fallen into decay, hillsides had been ravaged for fuel. People had suffered appalling hardships, and about one million had been forced back to China.

When the British Pacific Fleet steamed into Hong Kong on 30 August 1945 the harbour was littered with wrecks, and the land with derelict buildings. Commerce and shipping were at a standstill, and the population was virtually starving and without fuel. The military administration restored order and essential services and supplies, then relinquished its powers to the Hong Kong Government in May 1946.

Three events symbolized the war's end and a new beginning. On 30 August 1946 a cenotaph service honoured Hong Kong's war dead: the summer clouds were fleecy, and potted plants were in brilliant flower. Six months later, on 26 February 1947, the Japanese monument on Mount Cameron was destroyed: winter clouds hung low over the Peak when resounding cheers greeted its dynamiting. Sir Alexander Grantham, destined to be Hong Kong's Governor for a decade, arrived the following summer. In 1922, as a government cadet, Grantham had first seen the Colony in sparkling winter weather. Now, as his flying boat splashed down in July 1947, rain was falling. As Sir Alexander wrote:

> *The Peak and the 'Nine Dragons' were visible only intermittently through misty vapours that swirled around them. With the dark sea and sombre colours, the scene had something of a Wagnerian atmosphere.*

Sir Alexander's was an apt arrival. For, in the coming years, Hong Kong was surely to face 'Wagnerian' trials. Grantham was struck by the harbour's empty desolation: that was soon to be replaced with incessant activity, as the figures below for 1947 to 1948 suggest.

The immediate post-war population had been 600,000, but by 1947 it was already 1,800,000; and the excess of local births over deaths that year was over 40,000. Some 90,000 people were using the Star Ferry each day, three times the pre-war figure; civilian air traffic had more than doubled in a year; and there were already almost 12,000 vehicles, double the pre-war number. The wartime destruction of housing had mostly been made good; but the pre-war housing had been insufficient after 1937, and the population was greater still by 1947. In 1947 the Hong Kong reservoirs' total capacity was 27 million cubic metres, but, without rationing, the annual water consumption was almost twice that.

Freed from internment, the naturalist G. A. C. Herklots and his wife returned to their Pok Fu Lam home. They found only the shell of a building and every tree in the garden hacked down, including some ancient Chinese Banyans, Cotton Trees, and Bauhinias.

One of the most pressing tasks facing the civilian government in 1947 was restoring the hillside trees lost during the war. Post-war aerial photographs show that, though many *fung shui* woods had survived with their upper canopies intact, the reforested hillsides had been severely denuded. The higher slopes still had a cover of shrubs, but very few trees. The lower slopes were scarred red-brown where the tree-cutting had been worst. With summer rains hastening erosion, the reservoirs were in danger of silting up—as a 1948 government report observed:

> *From 1937 onwards, severe inroads have been made into the Colony's wood reserves, at first by illicit tree-cutting on the part of the swarms of refugees who fled into the Colony after the Japanese occupation of Canton, then by the fellings to make good the deficiency in supplies of firewood caused by the Sino–Japanese war, later by the Japanese to provide fuel during the occupation, and finally by the British military administration immediately after the war.*

Hong Kong's hillsides had been severely denuded during the war. This ridge, jutting into Tai Lam Chung Reservoir—completed in 1957—shows stark evidence of previous erosion.

Hillside fires were a constant menace during the 1950s and 1960s, due to the large population of poor squatters.

Implementing plans often thought out during their wartime captivity, government botanists and foresters experimented with new planting methods and species. By 1948 they were planting out each year some 100,000 Chinese Pine seedlings, and losing only 10 per cent. The hillsides were now burnt beforehand, to provide ash fertilizer and to reduce competing growth, and the seedlings were tended for some months after planting. Special attention was given to treeless areas around the reservoirs.

The Great African Land Snail, native to East Africa, had spread through Asia during the 1920s and 1930s. First seen in Hong Kong in April 1941, it had infested Hong Kong Island by the war's end. Voracious and long-lived, the snails did great damage to vegetation. During 1946 and 1947 forestry workers killed 236,302 snails (so the records state); and during subsequent summers Herklots regularly collected 'a bucketful before breakfast', then buried their remains under his new trees. Austin Coates regularly killed over 100 in his garden early on summer mornings—'huge armadas which advanced majestically upon all cultivated matter'.

The Great African Snail invasion paled by comparison with the post-war human onslaught. The refugees who flooded out from China after 1949 fled in fear, desperation, and mostly abject poverty. Faced with the depredations of the People's Liberation Army, and following the Communists' final victory in 1949, Hong Kong seemed the only hope. The government report 'A Problem of People', published in 1956, recalled Hong Kong's historic role as a refugee asylum, and the community's generosity towards refugees. As the report stated, during 'China's spasmodic but long-drawn civil war ... one of the few stable factors was the accessibility and availability of Hong Kong as a refuge'.

The statistics are stark. In about 1950, of Hong Kong's total land area of 1,076 square kilometres, only about one-fifth was 'usable' for agriculture or housing—and of this a mere thirty-odd square kilometres were 'developed'. The rest was too steep or swampy to build on without massive cost. Yet onto this scrap of land about one million people descended between 1949 and 1956: the vast majority of them were refugees, and very few ever returned to China. Hong Kong had always prided itself on its open border with

China, but in April 1950 a quota system was introduced to limit arrivals. By then the population was 2,360,000; by 1956 it was over 2,500,000, and increasing by about 75,000 each year. New Zealand, with over 250 times the total land area, and far more of it usable, had fewer people.

This influx of people transformed Hong Kong's agenda for two decades—and severely affected its natural environment. Never before had the Colony faced such relentless human pressure, never before had its space been so crowded or its services so strained. As 'A Problem of People' observed:

> *When one reads of one million homeless exiles all human compassion baulks.... Eventually the last vestiges of hope are centred on the calculating machine and the drawing board.*

The urban areas were already desperately overcrowded, and the poor were crammed into tenements and ramshackle temporary structures. Many refugee families had nowhere to live but alleys, stairways, and rooftops. Then, as 'A Problem of People' describes, 'they moved up the hillsides and colonized the ravines and slopes which were too steep for normal development'. There, living close to the urban areas, they could beg, borrow, or steal materials for a shack: cardboard, hessian, flattened tins, galvanized iron, even timber. And, close to the higher hills, they could forage for fuel: grass, shrubs, and trees.

An American writer, Christopher Rand, returned to Hong Kong from China in June 1951. Coming into the harbour Rand was astounded at the shanty towns spreading up over the Island's and Kowloon's summer-green hills: 'The flimsy little weather-beaten huts fascinated our navigator. "Extraordinary," he said, "They look like crates." They did.' Another newcomer wrote of hillsides 'with thousands of flimsy shacks perched on their steep granite faces, squatter settlements flowing down the mountainside like glaciers of rubbish'.

The squatter hillsides were crowded beyond endurance, with up to 5,000 people per hectare (in Australia, an equivalent urban area held about ten family homes and gardens). Sewage seeped down the slopes, and chickens, ducks, and pigs added to the squalor. Health and safety were major concerns. As Han Suyin wrote in *A Many Splendoured Thing*:

> *[The refugees' shacks] clung to the crumbling hill slopes, huddling beneath large threatening boulders, in danger of being washed away by the rains, in danger of being pulled down for health's sake, in danger of fire every time a meal [was] cooked.*

The treeless hillsides were prone to collapse after heavy rain. But winters, not summers, brought the greatest single threat: fire. Charcoal or wood 'chatties' (cooking braziers), candles, kerosene lamps, illegal electrical wiring—all posed constant hazards, especially in the dry months. Clearing firebreaks was almost impossible, as doing so created more homelessness: 'A match and a light breeze were sufficient to doom these inaccessible communities', warned a government report.

Finally the inevitable happened. A large squatter estate, at Shek Kip Mei in north Kowloon, caught fire: photographs show its people watching in horror as their community was destroyed. Beneath black billowing smoke, a flaming inferno leapt up the hillside—consuming whole rows of shacks in minutes—leaving only their blackened shells. So intense was the conflagration that its glow was visible even from the outlying islands. It was Christmas Eve 1953: when the flames died out early the next morning 53,000 people had lost their homes. Miraculously only two people were killed.

Tragic as it was, is the Shek Kip Mei fire relevant to this 'environmental history' of Hong Kong's 'natural landscape'? Surprisingly perhaps, it is central to the story—for four reasons.

First, Hong Kong's challenging climate and terrain had, in part, moulded its communal psyche. This led the community and government to respond to Shek Kip Mei not with despair but with the generosity and stoic resolve born of prior trials. Guangdong also responded generously.

Second, it was Hong Kong's natural lack of flat land that had prevented the government from building temporary refugee accommodation—or at least preventing such clearly dangerous concentrations of people as at Shek Kip Mei. Ironically, now there was some vacant land: a week after the fire bulldozers began levelling the blackened hillside to make way for 10,000 housing units. As 'A Problem of People' stated:

> *On Christmas Night, 1953, land was provided by an act of God. Forty-five acres were cleared of human habitation by the most extensive fire in the Colony's history.*

Third, since 1949 it had been Hong Kong's fervent hope that the refugees would finally return to China: that they had no intention of doing so was underlined by Shek Kip Mei. Hence, 'a temporary influx' became 'the permanent population'—one so large, for so small a place, that over the coming years it placed immense pressure on both the urban and rural environments.

Fourth, the government's acceptance of these people was symbolized—and made real—in its commitment, delivered soon after the fire, to rehouse at public expense the entire refugee and squatter population. This radical decision, and the overwhelming human and financial challenges it entailed, was made because 'the spectacle of such extremes of misery, need, and danger in the heart of a prosperous community could no longer be tolerated.' As 'A Problem of People' added with justice: 'the people of Hong Kong have pledged a portion of their own future for the benefit of strangers who took refuge here'.

'Building land in Hong Kong is not found, it is made; either hacked out of the hills or reclaimed', the report observed. Given the critical shortage of usable land, rehousing the population demanded 'vertical development', and, initially, cramped and spartan box-like flats. It was with some pride that, less than a year after Shek Kip Mei, the government could report:

> *In the spring we were building to two stories, in the summer to six and in the autumn to seven. In the winter we were preparing to demolish some of the two-storey blocks to make way for foundations capable of holding seven stories.*

Thus it continued: ever-taller 'resettlement estates' rose on either side of the harbour. Only photographs can give a true sense of the speed and extent of the rehousing programme—or of how hills were simply quarried into the ground to create space for housing. The building was not all low-cost housing. Around Hong Kong the higher hills were terraced for luxury apartments; and, especially behind Kowloon, factories and cottage industries spread ever further into the foothills.

Rapidly expanding industries provided the employment that gave the refugees their living—and yielded the revenue to finance the housing programme, and much else that the new Hong Kong needed: hospitals, schools, tertiary institutions, offices, reservoirs, transport facilities, even cemeteries. All consumed scarce land, all created waste and degradation in some form—but all provided the incalculable benefits of decent lives.

Hong Kong had virtually no natural resources, excluding its harbour and its people. Yet, between 1955 and 1970: the number of hospital beds per thousand people increased more than three-fold; on average one school for about 900 children was completed every week, year after year; industrial employment grew six-fold; and by 1970 average earnings were the second highest in Asia.

No transformation on such scale or at such a speed could have come without costs. Perhaps inevitably, as the government fought to meet pressing month-to-month and year-to-year needs, longer-term issues were pushed aside. The environment, in particular, was often neglected—and increasingly it suffered.

Hong Kong's weather took no notice of human affairs and needs. Had it been a normal year, the summer of 1954 would have brought some 2,000 mm of rainfall. Instead, only about half that fell—and, until March the following year, Hong Kong endured a protracted drought. The reservoirs almost dried up; the water supply was restricted to four hours a day or less; and further large squatter fires occurred as the shacks became tinder dry. A fire at Tai Hang Tung in July 1954 left 24,000 homeless.

The 1954 government report noted: 'A daily domestic routine of filling baths and buckets with water has become a regular feature of the life of rich and poor alike, and it is borne with an equanimity which surprises visitors.' The worst hardship fell on those living in old tenements or in hillside shacks, long queues of whom were constantly to be seen waiting at public standpipes. A government official wrote of these people:

> *They will be there in any weather; bitter cold or stormy heat, even in a downpour of rain. When their turn has come, each one will move off, stooping a little, and with hesitant steps under the weight of two kerosene tins of water suspended from each end of a bamboo pole.*

After meagre showers of rain, squatters living furthest from standpipes trudged up into the higher hills, then scrambled up gullies to where water trickled from streams and springs.

Hong Kong's three driest summers on record were followed, in May 1957, by the wettest May since the deluge of May 1889. Almost 900 mm of rain, three times the normal, was recorded at the Observatory that month—and at Beacon Hill over 700 mm fell on a single day (21 May). The rain brought silt and chaos to the urban areas—and terror to the squatter hillsides. During rainstorms, amid lightning and thunder, water cascaded down the hillsides in torrents of mud and stone, undermining precariously built shacks and threatening to bring boulders crashing down.

Late in 1957 Tai Lam Chung Reservoir, the first major post-war dam, began delivering water. When full it held over 20 million cubic metres, half as much again as Shing Mun Reservoir. In one of his last duties as Governor, Sir Alexander Grantham officially opened the reservoir on 7 December 1957.

Three weeks later, on Saturday 28 December, 25,000 people gathered at a sports stadium to bid farewell to Sir Alexander and Lady Grantham, who together had encouraged countless aspects of Hong Kong's post-war reconstruction. Sir T. N. Chau affirmed during an affectionate and moving ceremony: 'Your departure will mark the end of a decade which, in terms of material progress and the development of a community spirit, is unprecedented in the annals of the Colony.'

Even close to Victoria, steep hillsides and inaccessible gullies—such as this damp, mossy place—provide habitats for small animals.

It was a fine winter's day. At Kai Tak dust no doubt billowed up where an airstrip was being reclaimed from the harbour; work there went on seven days a week, much of it still by man-power. Across the harbour the new Central Reclamation had recently been completed, and the new Star Ferry Terminals had just been opened. Indeed, wherever one looked, urban development was pressing on apace, reaching ever higher, and spreading ever further afield into the country.

And, at a few places in the mountains—notably Tai Mo Shan and Ma On Shan—camellias were in bloom. One rare species, with cream petals and yellow-gold stamens, had the largest flowers of any camellia in the world. First discovered in 1955 by the forester C. P. Lau, it had been named Grantham's Camellia.*

* *Sir Alexander, mirroring the thoughts of many Hong Kong expatriates, wrote of their departure: 'Were we going home, or were we leaving home?'*

Photographs give clear evidence of Hong Kong's hillside vegetation in the early 1950s. The lower Kowloon hills were still almost entirely bare, their ridges ribbed by erosion; the only trees were in steep ravines. The hills high above Kowloon, and through much of the New Territories, had only grassland. Hong Kong Island's lower slopes were mostly similarly stark. Around Tai Tam the hills had a cover of shrubs, but very few trees.

New forestry schemes, more ambitious than any before, had been initiated in 1954. Almost 300 hectares of hillside were planted that year, mostly with Chinese Pines and eucalypts. Similar areas were planted in later years.

Besides minimizing erosion, especially around reservoirs, the post-1954 reforestation aimed to improve the lives—and incomes—of poor rural villagers. Government nurseries were greatly expanded, both to supply reservoir plantations and to provide seeds and seedlings to villagers. The country people still mostly used grass and wood for fuel, though they now accepted the need to conserve and plant trees. However, few would have done so, as Austin Coates notes, 'without encouragement and without

Hong Kong is today far better forested than in the 1960s. Photographs show that much of Cheung Chau, including this now well-wooded area, was then virtually treeless.

the knowledge that the administration would protect [their traditional hillside fuel] rights'.

Accidental fires and illegal wood-cutting, both usually at their worst in winter, still worked against the foresters. Urban squatters still scoured the uplands in search of fuel. Grass-cutters trudging homewards with heavy loads remained relatively common well into the 1960s. Thereafter, as villagers lost interest in exercising their traditional fuel-gathering rights, Hong Kong's reforestation became increasingly well established.

During the 1950s and 1960s, despite the urban overcrowding and squatter settlements, even close to Victoria and Kowloon there were still pockets where nature ruled. Further afield the rugged terrain had preserved large areas of countryside. As the post-war reforestation slowly spread, these places became increasingly appealing to small animals.

Christopher Rand, the American writer, was struck by Hong Kong Island's reservoir catchment channels: 'In wild parts of the Island you come on great works of masonry, looking amid the jungle like Renaissance engravings of what the Romans left.' He often hiked around the Peak, where dampness kept the hillsides lush. There, Rand wrote, one saw: 'bamboos, vines, flame trees, palms, giant tree-ferns; bananas with their vast leaves and sometimes dangling obscene flowers; brackens, mosses; hibiscus, azaleas, rhododendrons, ginger, iris, gardenias....'

Around Pok Fu Lam, in the early 1960s, the poet Edmund Blunden revelled in 'wooded and verduous hills'. For him the University of Hong Kong's grounds, set on the lower slopes of Victoria Peak, were the haunt of birds and butterflies, with 'frogs and grass-snakes and cicadas at their summer season'.

Small creatures—doves and dragonflies, bats and owls—enriched the Blundens' family life. Nature had barely changed since Europeans first came to Hong Kong, and indeed long before that—as Blunden wrote in his poem 'A Hong Kong House':

The handy lizard and quite nimble toad
Had courage often to explore
Our large abode.
The infant lizard whipped across the wall
To his own objects; how to slide like him
Along the upright plane and never fall,
Ascribe to Eastern whim.
The winged ants flocked to our lamp, and shed
Their petally wings, and walked and crept instead.

Birds slowly recolonized the new woodlands, among them Black-eared Kites and Blue Magpies. Snakes were endlessly being gossiped about. But Christopher Rand, like most walkers, saw very few of them—which was hardly surprising, since almost all of the local snakes are timid and nocturnal. Though only eight land snakes are harmful to man, and Crab-eating Mongooses are native to Hong Kong, there was talk of importing mongooses to prey on snakes. Rand commented of the mongoose plan: 'No one knows what tinkering with the natural balance might lead to'. Herklots was scathing:

Every living snake in your garden or basement is a venomous cobra; most dead snakes are harmless rat-snakes or grass-snakes. Every living snake is at least six feet long; most dead snakes prove to be three feet or even less.

There is little evidence concerning the populations of larger fauna such as Barking Deer and Wild Boar. However, Barking Deer still roamed in small numbers, even on Hong Kong Island: their barks sometimes echoed across wooded valleys, their droppings were seen quite often, but the animals themselves—also timid and nocturnal—were rarely encountered. Someone who did come face to face with one was Joyce Booth. From her verandah she had often heard deer; and, bewitched, she sat one night by a valley stream, surrounded by boulders and 'black, liquid darkness'. The only sounds were the crickets and cicadas. Then a male deer came down to water:

Through the underbrush the sleek head
Pushed its way,
Ears twitched and black nose shone
Like wet tar.
He bent his head to drink the water
And I saw his two, small horns,
His slanting eyes and glossy back.
I moved a fraction—no more—
But he got wind of me.
His head jerked up and tossed the water drops
Like silver spray
With startled leap he cleared the stream
And vanished in the trees.

Few people knew more about Hong Kong's wildlife than John Romer. Romer, a self-taught naturalist, was employed by the government soon after the war to control rodents—and, when he retired in 1979, he left a legacy of passionate knowledge about Hong Kong's fauna. In 1952, during a visit to Lamma Island to count fleas on the bats living in caves, it was Romer who discovered the tiniest of frogs: *Philautus romeri*, well known today as Romer's Tree Frog.

John Romer fought tirelessly against medicinal and gourmet customs, which saw the widespread killing of many endangered species—both here and in China. And he spoke out against the local obsession with money and the ecological indifference of his employers, the government. Given Hong Kong's present environmental woes, another nature lover's judgement of Romer bears quoting in full:

> *He was an ecologist when all too many people did not know or care what ecology was. If those in positions of influence and power had listened and acted, instead of merely studying the bottom line, Hong Kong would have been substantially free of present day horrors like Tolo Harbour. Too little too late could be a cry from [Romer's] grave.**

In the early 1950s, as the District Officer Austin Coates knew, 'The greater part of the colony was rural. One hour by launch or car from the hub of the city, and you could find yourself, if you knew where to go, in Chinese countryside.' Indeed, given China's post-1949 politicized agriculture, the New Territories were more traditional than China proper, as in rural Hong Kong land ownership and traditions were still rooted in ancestry and clan lineage, not Communist ideology. One could still see water buffalo, farmers working irrigation channels, and Hakka grass-cutters.

Even in the mid-1960s much of the agricultural landscape seemed virtually unchanged, as photographs taken then appear to show. But such images were often selective, nostalgic views: for by then the New Territories were being changed fundamentally.

Rice was still the main crop in the early 1950s, as one 1953 weather summary suggests: 'July was exceptionally dry, and favourable for the first paddy harvest'. In 1954, of Hong Kong's 70-odd square kilometres of arable land, 70 per cent was under 'wet' rice, and a mere 6 per cent was growing vegetables. Fifteen years later, in 1969, the rice area had shrunk to some 40 per cent, and more profitable vegetables now covered almost 30 per cent of the land. Previously farmers had kept pigs and chickens only incidentally, but by 1969 rearing them had become important. By then remittances formed a large part of rural incomes: it was easier to earn city wages, in Hong Kong or overseas, than to toil as a farmer.

The shift out of rice and into vegetables was due in part to the 1950s refugees, many of whom were Hakka market gardeners. But the changes were also driven by the new Agriculture Department—and by the visionary Kadoorie Farm and Botanic Gardens. By grafting agricultural science onto Chinese farming practices, and by improving rural water supply, access, transport, marketing, and finance, these two bodies assisted destitute refugee farmers to improve their lives. It was essential work, for by the 1950s Hong Kong's agricultural land was supporting some 300,000 people—or about five times the farming population of fifty years before.

Now known as Kadoorie Farm and Botanic Gardens, the Kadoorie Agricultural Aid Association Experimental and Extension Farm and Botanical Gardens was formed in September 1951 by the brother philanthropists, the late Lord Kadoorie and the late Sir Horace Kadoorie. Sited near the northern slopes of Tai Mo Shan, Kadoorie Farm's initial aim was to help poor refugees become livestock and cash crop producers. Since the 1970s the Farm has concentrated more on botanical and ecological research, and is now a key contributor to Hong Kong's ecological knowledge.

Better rural incomes came at a cost. The New Territories of the early 1950s recycled virtually all its waste as manure. But by the mid-1960s the huge increase in the rural population, as well as in chickens and pigs, meant there was far more human and animal waste than agriculture—or nature—could absorb. The results were not evident for another decade: but by the late 1960s the streams of the agricultural lowlands were steadily fouling. Some were already virtual sewers.

The Tanka fishing people were impoverished by 1945. After the war the government and the new Fish Marketing Organization, a cooperative, worked to revitalize the fleet. During the 1950s and 1960s the junk fleet's mechanization led to much greater catches, and more prosperous lives for the Tanka. However, over the longer term, inshore fish stocks were depleted due to the boats being able to fish in almost any weather; and new synthetic materials led to much greater amounts of marine rubbishing. Meanwhile, the increasing volumes of human and industrial wastes began degrading the coastal waters—which until the early 1960s had been largely unpolluted.

'You have the sense of making a journey through a doomed landscape', F. D. Ommanney, a marine biologist, said of the New Territories in the early 1960s. Superficially rural life still seemed traditional—'people straight out of a Chinese painting done many centuries before you, or your Hillman Minx, were ever thought of', he wrote. In fact the country was threatened as never before by the city's inexorable expansion and industrialization.

No one knew this better than the District Officers of the 1950s and 1960s. Charged with the welfare of Hong Kong's villagers, they knew that the New Territories' rural life was surviving on borrowed time. Austin Coates was responsible in the 1950s for the

* *Graham Reels, an ecologist at Kadoorie Farm, showed me this material on Romer. It comes from a paper by Joe Fearon, published by the Hong Kong Natural History Society (1988).*

people of Tsuen Wan, then an industrial yet still agricultural area north-west of Kowloon:

> *My duty was to fight a rear-guard action against urban encroachment, and to protect agriculture and village life, wherever this was desirable and possible, in order that the country people should not suffer by too rapid social and economic changes.... Somehow every acre of cultivated land had to be kept in agricultural use for as long as possible, to allow time for the original inhabitants to adapt themselves to the new environment that had quite suddenly installed itself around them.*

Fung shui still patterned the land. But 'capital outlays' and 'returns on investment' were driving Hong Kong towards its industrial future.

More than in any other period, creating an adequate water supply was perhaps Hong Kong's greatest environmental challenge in the 1960s. During the nineteenth century the water problem stemmed largely from the Territory's steep terrain, the resulting lack of natural storages, and the monsoon climate. But long before 1960 the overriding issue that made the water supply a constant, nagging problem was Hong Kong's sheer number of people. The 1960 population was almost three million; it was predicted to exceed four million by 1970; and industrial water consumption was increasing rapidly. Projections of future water consumption were very difficult, as current use was based almost entirely on rationed consumption.

During the mid-1950s and early 1960s droughts threatened Hong Kong's existence. Shing Mun Reservoir, shown here before the summer rains, was often even emptier then.

During 1960 Hong Kong enjoyed unrestricted water use for only thirty-five days. The government annual report for that year opened with a lengthy chapter entitled 'Hong Kong's Water Supplies—1960, A Year of Decision'. Tai Lam Chung Reservoir, completed in 1957, had vastly improved the situation; in 1960 China agreed to provide water from the Shenzhen Reservoir; and the Shek Pik Reservoir on Lantau Island was due to be completed in 1964. None the less, as the commentator Gene Gleason wrote in 1964:

> *If the government builds all the reservoirs the colony needs, who will pay for them? If it doesn't, how can the fast-growing population of the colony survive?*

The Hong Kong solution was to turn topography on its head by transforming a sea inlet into a freshwater lake. Hence Plover Cove Reservoir was born: by enclosing Plover Cove with long dams, a gigantic water storage could be built that would triple the entire existing capacity. Moreover, Plover Cove was to be linked to an integrated maze of pipes, tunnels, and filtration units which could supply water to the urban areas destined to soon spread into the countryside.

Yet despite such schemes water consumption remained a constant concern. Government water engineers made their plans in ignorance of the *actual* future population and weather—both of which dealt further blows to Hong Kong's environment in the early 1960s.

In China, the Great Leap Forward (initiated in 1958), the forced collectivization of agriculture, the drastic increase in urban populations, and various natural disasters—all combined to precipitate what is said to be the worst famine in human history. It is estimated that, between 1959 and 1961, about thirty million people died—and once again refugees swarmed into Hong Kong. In May 1962 alone, over 50,000 crossed the border, though most were apprehended and returned to Guangdong.

Two severe typhoons brought much needed water—and widespread destruction. In June 1960 Typhoon Mary, the worst since 1937, caused great damage. A far worse typhoon struck on 1 September 1962: Typhoon Wanda. Wanda passed very close to Hong Kong that morning, producing winds gusting over 250 kilometres per hour. Fifteen ocean-going ships were driven ashore, some 500 fishing junks were sunk, and upwards of 10,000 people were made homeless. About 170 people were killed or missing; had the typhoon not hit in daylight, the toll would have been much greater.

Potentially far more serious, drought conditions began to affect Hong Kong from October 1962. The first six months of 1963 were the driest on record: the June rainfall, just over 200 mm, was half the normal—but almost torrential compared with the previous seven months' total of about 70 mm. By then the earth was cracked; the grasses had turned brittle; and the reservoirs were virtually empty.

Hong Kong was facing a crisis. The water supply was cut to four hours once every four days. Water normally piped from Shenzhen across the border was much reduced, because of drought conditions there. Hong Kong was forced to rely on tankers bringing water from the Pearl River. These ships made some 600 voyages, and brought about eight million cubic metres of water to Hong Kong—or about fifty-six days' heavily rationed supply. The emergency lasted until the late summer of 1964, when normal summer rain fell at last.

Two years later a freak rainstorm inundated the territory. On Sunday 12 June 1966 Hong Kong awoke to one of the most concentrated downpours ever witnessed here, when in a single hour over 150 mm of rain was recorded in places—and almost three times that in twenty-four hours. The hillsides were slashed with landslides; the urban areas were awash with silt and debris. Many people were killed and thousands made homeless.

Three years later, in August 1969, consistent heavy rain brought the reservoirs to their maximum capacity. Plover Cove Reservoir, opened the previous January, overflowed for the first time on 11 August 1969, thus promising a more certain future.

The Hong Kong of the late 1960s offered, for most people, an infinitely better life than in the immediate post-war years. The Territory was increasingly prosperous, and its long-term survival was no longer in doubt. But the Colony's transformation had come at a human cost.

Incessant toil had reshaped Hong Kong's physical environment. For most people, the post-war decades had brought only work—recuperation—then more grinding hours. For most, a day off was an annual luxury, and a bus fare too valuable to spend on a trip to the beach—let alone to the more distant New Territories.

The riots of 1967 were instigated by Communist agitators, but the unrest of that torrid summer did also reflect a wider disquiet. Hong Kong people now had a vision, and expressed the demand, for better living conditions. These the next decades would mostly bring.

Meanwhile, insidiously and initially almost unseen, the natural landscape was being steadily degraded. By the late 1960s various combined factors were speeding Hong Kong's environmental decline: a refugee-born, transient, communal psyche; the colonial mentality of some decision-makers; rising affluence and technological power; inadequate government controls; and sheer greed. Together these spawned pollution and other environmental ills which, though mostly originating in the urban areas, were soon to affect the countyside.

In the years just after the war Graham Heywood, still a keen hiker and by then Director of the Royal Observatory, had observed:

> *Hong Kong has an incomparable setting, but hitherto man has done lamentably little to contribute to the natural beauties of the place. The city sprawls untidily, with few green spaces or noble buildings to grace it, and there is a pressing danger that much of the surrounding country may be spoilt by unsightly building developments.*

By the late 1960s the impact of urban development on the surrounding countryside was infinitely greater. *The Atlas of Hong Kong,* a cheap school atlas of the late 1960s, a product of The Genuine Book and Stationery Company, observed more colourfully—and with both admiration and awe:

> *The mushroom growth of Hong Kong is rapid in recent years, and it is now an enchanted land with skyscrapers, theatres, nightclubs, and limousines.... But inwardly there are problems and problems, latent and manifest, which could only be solved by a very careful plan. All these, of course, are under the Government's consideration for the coming decade.*

The massive Shek Pik Reservoir, seen here below Lantau Peak symbolized Hong Kong's achievements in the 1960s—and promised a better future.

Grand vistas typify southern Lantau. This view from Fan Lau Kok, the island's southern extremity, looks north-east towards Lantau Peak.

REGION

LANTAU ISLAND

The farmers complain that the Water Engineer does not fully understand the needs of paddy farmers. The Water Engineer's position is that the farmers are being selfish, that the needs of the majority of the urban people of Hong Kong and Kowloon must take precedence over the villagers of Tai Yu Shan [Lantau]—and that Progress will eventually help the villagers.

ARMANDO DA SILVA, 1972

It is autumn. I am camped on a ridge near Lantau Peak, high above the Po Lin Monastery. Around me the uplands of Tai Yu Shan—Lantau—are as empty and wild as ever: a craggy landscape that rears up in striking grandeur and stretches out of sight past serried peaks.*

But in the valleys below man has transformed the natural scene. To the south, sheer slopes fall towards the Shek Pik valley, where today a reservoir forms a picturesque expanse. To the north gullies drop down towards Tung Chung valley, once a peaceful haven but now part of the awesome airport developments around the nearby island of Chek Lap Kok.

A generation or so has passed since the Shek Pik Reservoir was built. It was in the early 1960s that Armando da Silva, quoted above, heard the views of the villagers living near Shek Pik. They were distressed about the reservoir then being constructed on 'their' island, which would inevitably disrupt their paddy water supplies.

Is it only a generation since Hong Kong's need for reservoirs impacted on Lantau's landscape—since its rural life was challenged by the problems and values of the urban population? Can it really be a mere generation since a Lantau villager wrote on 9 March 1961 to his District Officer:

Sir,
The hillside area behind my hut is known, in fung shui *terminology, as the Dragon's Vein, and is therefore of great importance to our villagers.*

This fact notwithstanding, an outsider has had the audacity to hire some workmen to dig up the earth there in an attempt to build a house on the site. In so doing, he has neither obtained the consent of the village elders nor applied to your Office first for a survey. Thus no sooner had the work started than the village's livestock, such as cattle, pigs and dogs, were afflicted with disease and ceased to drink or eat.

Their condition has shown some slight improvement only after I had the holes filled up, and after a charm was employed to invoke the gods to drive away the evil spirit....

*[signed] Representative of Shap Long Village, Lantau.**

Long before the name Hong Kong had any significance, Tai Yu Shan (Big Island Mountain) was known to Chinese officials, traders, and scholars. This was partly due to its sheer size. Dominating, majestic, it rose above the eastern estuary of the Pearl River, south of Guangzhou. But through successive Chinese dynasties the island, absorbed in its own fishing-farming rhythms, stayed apart from mainland events.

By the Song Dynasty, Punti and Tanka clans were well established on Lantau, and strong enough to resist imperial demands for tax on their produce. And by 1277, when the two Song boy emperors fled to Tai Yu Shan as their dynasty finally collapsed, the island was much changed. By then the activities of fishermen, farmers, and pearl divers—and the production of salt, joss, and lime—had significantly denuded the island. After the coastal evacuation of 1662 Hakka people also settled there.

* *The name 'Lantau' is generally used here. Though other translations are sometimes given, 'Big Island Mountain' is true to the characters for Tai Yu Shan.*

* *Lantau's District Officer was then James Hayes, who fortunately preserved this telling letter.*

These denuded hills in southern Lantau bear witness to previous deforestation by villagers. The grassed-over hollows show the sites of old landslides

Yet even when Lantau became part of the Colony of Hong Kong in 1898 it remained largely a world unto itself. 'Lantau', a colloquial Cantonese name meaning 'Cleft Skull' came into use then with Europeans. But to the island's villagers it remained Tai Yu Shan, as it does today.

A ferry service began from Victoria to Mui Wo in 1938, but not until about 1960 was Lantau drawn into the modern world. Then 'progress' changed the island forever: when the Shek Pik Reservoir was built—then the most dramatic alteration to the island's physical environment caused by man.

Today, along north-western Lantau, at Tung Chung and nearby Chek Lap Kok, the developments for Hong Kong's future airport are affecting the island far more drastically than Shek Pik ever did. But—at least for the present—almost all of middle and southern Lantau is still largely wild. Lantau North Country Park and Lantau South Country Park together cover almost 8,000 hectares, and the Lantau Trail winds for 70 kilometres through their rugged uplands.

Describing the Lantau of early this century, G. R. Sayer wrote of 'mysterious peaks' amid country whose 'unaffected beauty provided the perfect foil to Victoria's more cultured charms'. Lantau still has abundant natural grandeur: from Mui Wo, around the Chi Ma Wan Peninsula; south along the coast to Shek Pik and Fan Lau; up to Ngong Ping and the Po Lin Monastery; and so along the exhilarating ridges near Lantau Peak and Sunset Peak.

Shaped like a short-legged dragon, Lantau is some 30 kilometres long and covers 142 square kilometres—or about twice the length and area of Hong Kong Island. Viewed from the side, Lantau appears to rise abruptly from the sea. In fact, the island is unusually steep (as its map contours show). Much of Lantau is over 400 metres, about the height of the Upper Peak Tram Station on Hong Kong Island—and its two highest peaks reach well over twice that.

The mountains, folded north-east to south-west, run like a backbone along a central spine. Plunging slopes and ravines descend to deep valleys. The streams are very short and, after heavy rainfall, violent. Northern Lantau is mostly granite but the southern peaks, higher and steeper than the northern ones, are the result of volcanic eruptions. Alluvium washed down from these volcanic uplands forms rich coastal valleys.

The Lantau we see today is dominated by grassy hills. As the World Wide Fund for Nature's 1993 *Hong Kong Vegetation Map* shows, nowhere else in Hong Kong is there such widespread 'grassland'—or, proportionately, so little 'woodland'. Newspapers and television screens showed the results of Lantau's lack of trees when, in November 1993, exceptionally heavy rainfall inundated the island. The hills were scarred red-brown where countless landslips occurred, and the lowlands flooded by silt-laden torrents.

Yet as late as the Song Dynasty, Lantau was still at least reasonably wooded. And when an ancestor of the Mo clan settled there about 1300, some twenty-four generations ago, there were sufficient trees for him to become a woodcutter. Like many other families', the Mo's 'rice bowl' came from making charcoal.

The need for domestic fuel and timber obviously diminished Lantau's woodlands, but probably the greatest deforestation was caused by tree-cutting for the island's lime and charcoal industries that supplied Guangdong and possibly other parts of China. The kilns, which produced lime from shells, date back to the Tang Dynasty. Each kiln load of shells must have used large amounts of timber and shrubs. The industry died out about AD 1000, but began again early last century and continued at least until the 1940s.

Grass-cutting and burning also hastened the replacement of woodland with grassland. *Mau tso*, 'thatching grass', was harvested regularly; and grasses and brushwood were always being cut for kindling. Grass-cutters often cut down saplings unintentionally, and taking grass away removed future compost and weakened the soils. Fires, though more immediately destructive, were probably less damaging than tree-cutting in the longer term—as the latter removed vegetation from the land, whereas the former transformed it into nutrient-rich ash. Trees cut for fuel, of course, were also burnt, but their ash fertilized farmers' fields, not the hillsides where the trees were cut.

The cumulative result was inevitable: except in inaccessible places, Lantau lost almost all its trees, and, especially where the slopes were steepest, its best, primordial soils. Grassland took their place. As Rudolf Krone, the German missionary, wrote in 1858:

> *The mountains [of Lantau] have usually a dreary and barren aspect, and resemble those of Hong Kong and the opposite mainland. The granite rocks are scantily covered with soil, and are overgrown with grass. A luxuriant underwood is found in the ravines, but trees are seldom met with, though groves of them, evidently planted, are generally found in the neighbourhood of villages, Buddhist monasteries and temples.*

Since the mid-1950s government reforestation has achieved considerable success, by establishing plantations on eroded areas in a few decades. However, inevitably on so large and mountainous an island, the Lantau plantations are mostly on the lower and middle slopes. Also they are not yet 'mature', and usually have just a few dominant exotic species such as Acacias, Casuarinas, and Pines—though more recent plantings include native species.

Deforestation gradually destroyed the habitats of Lantau's larger mammals and forest birds. Today, excluding a few Barking Deer and Wild Boar, there are no large native mammals. The commonly seen animals today are those adapted to grassland and scrub.

But, as almost everywhere in Hong Kong, butterflies are common. About a decade ago, Stella Thrower often saw twenty or more butterfly species during summer walks around the Chi Ma Wan Peninsula's reforested hills: 'great mormons, mottled migrants, common mimes, great orange tips, six rings, angled castors, skippers, common grass yellows and grass blues'.

Lantau's eastern coast stretches from the Chi Ma Wan Peninsula southwards towards Shek Pik—and from there on to Fan Lau, the island's southernmost extremity. From Chi Ma Wan to Fan Lau is 14 kilometres as the crow flies, but almost twice that around the bays and headlands.

Progress has its advantages: I confess always to taking the bus from Chi Ma Wan to Shek Pik. On one side Sunset Peak and Lantau Peak rise up, their massive flanks cut by gullies. On the other, just below the road, are some of Hong Kong's longest sand beaches. Out to sea—it would hardly be Hong Kong without some beckoning islands—are the Soko Islands and their satellite islets.

The southern tip of Lantau, Fan Lau (Division of Flows), is six kilometres from Shek Pik along a catchment path and later around a cliff-side track. Mostly following the contours about 100 metres above the sea, the route winds past craggy bluffs and coves.

The southern line demarcating Hong Kong and Chinese territorial waters formerly met with Lantau two kilometres north of Fan Lau: at Kau Ling Chung. There, in 1902, a small obelisk was erected by the men of HMS *Bramble*. Having scrambled up a sheer rocky slope, they set the marker high above the tideline 'to protect it from possible inroads of the sea'. Windswept and rarely visited, the marker is still there. Later, following an adjustment, the line

Lantau's south coast, at Fan Lau Tung Wan. To the right of this photograph mounds of rubbish, a common sight on remote beaches, presented a different mood.

Tai Yu Shan, today's Lantau, was well known long before Hong Kong. The remains of this fort at Fan Lau probably date back to the mid-seventeenth century.

was moved a few kilometres further south.

Fan Lau's villagers probably wondered why British officials bothered marking such a sterile spot: a headland no islander could ever use or farm, and which even the most intrusive government would never consider taxing.

Just to the south is Fan Lau Tung Wan, a crescent of sand backed by coastal scrub and abandoned paddy fields. When I was last there wintry waves washed in from a pale sea. Mounds of flotsam—the same mostly plastic mess that lines every Hong Kong foreshore too remote to be cleaned—marked the tideline. Should I photograph the waves or the rubbish, I wondered?

From Lantau's rocky south-westerly extremity, Fan Lau Kok (Division of Flows Point), on calm days in past years a sharply defined blue-brown line could be seen cutting across the water: the actual 'interface' of blue ocean water and brown river water. The line was visible when I was there in 1994. But, with the water muddied by dredging in the Lantau Channel the division has been blurred.*

Twice I have camped at Fan Lau, nestled in a grassy hollow near the Qing Dynasty fort there. Twice I have been astounded at the night-long stream of vessels passing around the point. Through the centuries this was always a busy trading route, but now the traffic is relentless: container lighters, sand barges, coasters, tugs—each one a cog in Hong Kong's and Guangdong's rapid economic growth.

'Growth'—might it one day destroy the timid Barking Deer still roaming Lantau, now so few in number that seeing one is exceptional? The last time I camped at Fan Lau a strange barking noise came through the darkness. I had never before heard a Barking Deer: was this one, or just a wild dog? The next morning I heard the same, somehow plaintive sound. Seconds later, a waist-high deer bolted past me—and was gone.

From Shek Pik (Rocky Wall) to Ngong Ping (High Plain) a trail winds up the south-western flank of Lantau Peak. About 500 metres above sea level, approaching the Ngong Ping plateau, one looks down into the bowl of Shek Pik's valley and reservoir.

Volcanoes long ago poured lava over the area. Explosive eruptions threw up rocks which dropped to the ground amid settling ash, so forming breccia (a fine-grained, extremely hard rock). The old lava can be seen today along the slopes south-west of Ngong Ping. The breccia, which has largely resisted erosion, forms the jagged scarps around Lantau Peak and Sunset Peak.

These volcanic remains look down on what appears to be a dead crater, its seaward side blown apart: the Shek Pik valley. But the valley, in fact, is the result of massive erosion.

Punti settlers established villages near Shek Pik sometime during the Ming Dynasty, and their dozen or so clans continued fishing and farming there for centuries. The population declined during the 1920s and 1930s, but the villages endured. Then, in a single decade, from 1950 to 1960, the needs of urban Hong Kong overwhelmed Shek Pik's fields and farms.

Three writers have left vivid impressions of Lantau during that decade of transition and upheaval: Christopher Rand, the American correspondent who lived at Ngong Ping during 1951;* James Hayes, Lantau's ever-enquiring District Officer from 1957 until 1962; and Armando da Silva, who did extensive field research there between 1962 and 1963. Together, they reveal a picture not only of Lantau but also of other, similarly remote parts of Hong Kong as they were then.

Well into the 1950s Lantau's villagers continued to live much as they had for generations. Their houses had fir-pole beams and rammed-earth floors; water came from streams and wells; malaria was common; and police patrols bringing basic medical care (and

* *One of the sites considered for a new power station, in fact later built at Black Point (near Tuen Mun), was the neck of land just behind Fan Lau Kok. The area, originally excluded from Lantau South Country Park because of a village, is surrounded by some of Lantau's grandest landscapes.*

* *Rand stayed at the farm being established at Ngong Ping by Brook Bernacchi, then a young lawyer experimenting with tea growing. Brook Bernacchi, now a leading lawyer, still has a home at Ngong Ping.*

gentian violet) were always welcome. Most villagers were illiterate. The women generally did the farm work, and the men did the fishing. As a Chinese proverb says,

> *When in the hills, live off the hills;*
> *When on water, live off water.*

'They seemed much like other South China peasants', Rand wrote. 'They grew rice, dug for shellfish, grazed little draught-cattle.' Charcoal making, hillside liquor distilling, stone cutting, and rice and vegetable farming continued through the 1950s, though with steadily fewer people involved. Fishing craft still relied on the wind, and natural fibre nets were spread to dry around the beaches. There was seaweed collecting and stake-net fishing, as Rand described:

> *The fishermen worked their spider-spread stilted nets, hauling on ropes to make them rise or fall.... Tiny men in tiny boats, as the Song painters had seen it.*

The rugged terrain still demanded sheer endurance. 'I never saw a wheeled conveyance on Lantau, not even a wheelbarrow', wrote Rand. Villagers hastened through the hills, often jog-trotting under heavy loads; the grass-cutters 'looked like moving haystacks'. The village paths were 'now level, now steep and flagstoned like dragons' backs, now gracefully curved, and at times appearing to hang out over thin air'.

At Lunar New Year villagers came down off the hills shouldering bamboo poles hung with lucky papers and boughs of small pink bell-shaped flowers, *tiu chung fa*—or Hanging Bell Flower. The species, *Enkianthus quinqueflorus*, had been protected by law since 1947. But how could the ban be enforced in Lantau's remote uplands? And who could begrudge villagers making some traditional New Year money from selling wild flowers?

During the dry winters hillside fires were common, 'filling the air with smoke and crackle'. Kites hovered around the blazes, pouncing on small animals fleeing the flames.

Summers transformed the landscape. Magnificent cumulus clouds rolled in—water ran down onto the fields—and the hills turned a rich green.

Government meteorologists measured *Lantau's* rainfall. *Tai Yu Shan's* villagers still relied on intuition and the *Farmers' Almanac*, based on complex Chinese calendars. The rice rains began with the Tuen Yeung festival in the fifth lunar month—and ended with the Chung Yeung festival in the ninth lunar month. The rainy season fortunately coincided with the heat that rice growing needed. But it also brought *dai fung*—typhoons that could flatten the paddy, strip grain off the stalks, and disrupt the water channels and terraces.

The Lantau villagers still inhabited an integrated world of hills and streams, fields and houses, clan lineage and *fung shui*. Special words, for example, described the practical and spiritual aspects of watercourses—as da Silva recorded. Streams called *kai* were intermittent upper streams flowing into *hang* streams, which flowed through steep valleys and always had water in their lower courses. Streams called *lung* flowed through level fields and were auspicious. These and other terms minutely defined both the streams and their relationship to the spiritual landscape. Similarly, coastal terms described different kinds of bays: with fresh water streams, with sandy beaches, with deep water—a *chung*, a *wan*, a *wo* (hence Tung Chung, Chi Ma Wan, Mui Wo).

Thus Lantau's village life—patterned by tradition and belief—endured through the 1950s, as the old ways did in other remote parts of Hong Kong too. But these rural worlds were surviving on borrowed time. Lantau's population barely changed from 1950 to 1960. But in that single decade Hong Kong's overall population grew exponentially, and there was rapid industrialization. Even before 1960 Hong Kong urgently needed another large reservoir.

Taro plants are common along valley streams, such as this watercourse in the hills above Mui Wo.

Waterfalls such as this one, near Mui Wo, once fed summer run-off into lower valley streams which Lantau's farms and villages depended on for water.

In 1961 the construction by outsiders of a mere house on Lantau village land could prompt bitter indignation—as the Shap Long villager's letter to James Hayes, quoted at the beginning of this chapter, suggests. The letter ended:

> *This man has no respect for our native traditions and is planning to tamper with the earth again. As this lawless character is not likely to show the least concern for our safety, would you please send an officer over as soon as possible to prevent him from carrying out these activities.*

Like most emotive pleas invoking *fung shui*, as James Hayes knows from first-hand experience, the letter probably reflected genuine personal and spiritual concerns, an insular (and clannish) dislike of outsiders—and a keen nose for compensation. The whiff of payments for government land resumptions had wafted over from Hong Kong and Kowloon during the 1950s.

If a house could provoke such a tirade, what dragons was the government ignoring—and abusing—in planning to build a reservoir at Shek Pik? Hayes observes: 'From the *fung shui* angle, Shek Pik was probably the worst place to construct a reservoir and so interfere with the local landscape on a really drastic and irrevocable scale.'

But valleys with favourable *fung shui* were generally also good for farming, and often ideal for reservoirs. No doubt inevitably, the government planners concluded that Shek Pik was the last large valley that *could* be flooded. Geological investigations had begun there in 1954, and in 1959 contracts were awarded for a reservoir to hold almost 25 million cubic metres (about four times Tai Tam Tuk's capacity). In 1959 Lantau's first vehicle road was built, from Mui Wo to the reservoir site. It broke down Shek Pik's isolation forever and, among other benefits and losses, allowed seriously ill villagers to be taken to hospital on Hong Kong Island. But what was a hospital? As a 1960 government report observed, before the road was built:

> *The villagers at the western end of Lantau were cut off almost entirely from the twentieth century.... Few activities of the government had penetrated the [Shek Pik] valley, and most of the inhabitants had probably never seen a railway train or a motor car.*

Though Shek Pik's clan leaders viewed the reservoir scheme with the utmost fear and hostility, it proceeded. Local people sometimes disrupted the development, especially when graves were affected and when a *fung shui* wood was destroyed. Then, in the early winter of 1960, they were 'relocated' from their ancestral lands—from 'their' valley, now a red-brown construction site. As the same government report recorded:

> *Last minute delays, caused by the gods choosing an unexpectedly late lucky day for the move, kept the village on Lantau until November, and 202 persons finally moved on the 22nd of that month. The oldest inhabitant was a lady of 86 who had never before left the island of Lantau. The contractors were waiting; the earth-moving machinery moved in and rapidly obliterated all signs of a village which had stood in the valley for about 600 years.*

Shek Pik Reservoir was opened on 28 November 1964. During 1963 and 1964 Hong Kong had endured its most critical drought, when its reservoirs were virtually empty and tap water was severely rationed. But the farmers in south Lantau had other worries. Reservoir catchment channels now encircled their hills and took water on to Shek Pik, and so, in neighbouring valleys, fields were being starved of water. *Kai* streams now fed into concrete channels, so the *hang* streams no longer flowed so well. No doubt genuinely sorry, the Water Engineer could only say: 'Progress will eventually help the villagers'.

Lantau Peak (934 metres), properly Fung Wong Shan or Phoenix Mountain, is Hong Kong's second highest mountain. From the Po Lin Monastery a boulder pathway winds up to its high spurs, past slopes that plunge down to Shek Pik. Below the summit a ceiling of cloud often closes in and, as vapours envelop the slopes, one's senses are heightened. The majestic panoramas disappear—the hills seem to shrink—and one is left looking at the ground.

On the sheltered northern slopes of Lantau Peak, in areas too inaccessible for past tree-cutting, there are woodlands with well-

Lush growth lines this cleft, 200 metres below the summit of Lantau Peak. The surrounding steep slopes sustain only scattered trees and shrubs.

Paths wind around the western side of Lantau Peak at about 600 metres; the track to the left goes up to Lantau Peak, that to the right descends to Shek Pik.

established native trees. In the ravines there is a rich diversity of species—of trees, shrubs, and bamboos. And in hollows kept moist by mists grow rare plants: ferns, orchids, and flowering shrubs. Lantau Peak and Sunset Peak, 'Sites of Special Scientific Interest', were so designated because of their flora.

One cannot but be struck by Lantau's ecological value—and by the sheer grandeur of its peaks, hills, and coasts. Yet of all the Hong Kong countryside, except perhaps Mai Po, Lantau is now the most severely threatened place. The view from Lantau Peak over Chek Lap Kok convinces one of that: for, spread out like some vast sandpit, is the awesome expanse of the airport construction site.

The future airport, necessary as it may be, has already destroyed the country aspect of Lantau's north-western coast. The new Lantau Port will only add to the loss. However, provided the airport and port can be kept completely insulated from the rest of Lantau, the destruction can stop there.

But were the airport and port to be linked by road to the rest of Lantau and to its existing communities, the island's natural landscape would be threatened as never before. The boundaries of Lantau South Country Park and Lantau North Country Park lie well inland, and thus large parts of coastal Lantau could easily be lost to widespread construction—leaving a fringe of residential and resort developments around the island.

To compensate for the loss of countryside that the airport required, the government promised to extend the boundary of North Lantau Country Park. Yet the actual extension, put forward annually by the Country Parks Board, has been refused each year. Why? Because, the government maintains each year, it lacks the necessary funds to finance the extension. The initial amount needed is about HK$20 million. Chek Lap Kok, as the government boasts, is the costliest public infrastructure project in the world.

The government's response demands challenging. Indeed, the views of some conservationists—who believe that the real reason for the government refusing the Country Park extension is that Crown land is being surreptiously set aside for development—must at the very least be examined seriously.

Lantau is no longer an isolated island of poor, uneducated villagers. Many people there share urban Hong Kong's values. Recently Lantau villagers intent on improving their television reception cut down a venerable *fung shui* Banyan. *Fung shui* may still matter to some—but money is god-like. Windfall financial bonanzas would inevitably result from connecting the island's communities to the airport or port (and so to urban Hong Kong). To countless Lantau people—and to numerous developers—such gains beckon like solid gold.

In 1951 Christopher Rand was troubled to see Lantau villagers coming down from the hills laden with wood. Rand asked the local police inspector what could be done. 'Tree-cutting is illegal but grass-cutting isn't', the policeman answered. 'Where do you draw the line between trees and grasses?' And Rand concluded, 'Why shouldn't a few poor souls make their living from illegal wood-cutting?'

The choices concerning Lantau today appear far more complex than Rand's wood-cutting dilemma. Yet in some ways they are far simpler. Reforestation can mostly transform barren hills, albeit very slowly. But wild countryside, once 'developed', is gone forever.

Plantation woodland spreading up the Shek Pik valley, about 100 metres above the Shek Pik Reservoir.

This view, from the higher slopes of Lantau Peak, looks west over the Ngong Ping plateau and on towards the summit of Cheung Shan—449 metres high.

By the mid-1980s almost all of Hong Kong's old agricultural lowlands had been 'developed'. Here, from The Hunch Backs, hills slope down to Sha Tin New Town.

HISTORY 1970–1990

COMMUTER COUNTRYSIDE

We have an advantage—which I have not seen equalled in the world—of magnificent areas of mountain, beach and island exceptionally close to the centres of population.

SIR MURRAY MACLEHOSE, 1972

Hong Kong's present approach to environmental protection is based essentially on the philosophy of 'disposal'.... This strategy ultimately calls for the use of the sea as a sink, the atmosphere as a sewer, and the land as a dumping ground.

K. C. LAM, 1983

Throughout Hong Kong's human history fires have been a feature of the natural landscape. Many were deliberately lit, some were accidental. But by the 1970s, and still more in the 1980s, the vast majority of countryside fires were caused by sheer carelessness: for by then the rural population had greatly declined, while the number of city people visiting the country was steadily increasing.

During the winter of 1983 and 1984 about 400 fires devastated some 5,000 hectares, or more than half the area of Hong Kong Island. Perhaps 200,000 trees were destroyed. Two years later the toll was still greater, as very dry winter months saw the immolation of about 500,000 trees. When darkness fell on 23 October 1985 no fewer than forty hillsides were blazing; most left behind black wastelands pock-marked by the charred shapes of countless graves.

Good rains the previous summer had brought on very tall grasses, and by mid-winter they were tinder dry. Through December there were numerous wildfires, and during January 1986 fire-fighters fought a string of countryside blazes. The worst began on 8 January, a conflagration that laid waste the Shing Mun valley and Tai Mo Shan—and threatened Kadoorie Farm.

The *Standard* judged that, as hill fires rarely brought loss of life, Hong Kong's highly urbanized people were blasé about them. Yet as the newspaper put it, 'every fire in a Country Park or a woodland is a potential ecological disaster'. The newspaper might well have been discussing Hong Kong's wider environmental woes when it concluded:

Fire-fighters need the cooperation of the Hong Kong people. Will we, the public, continue to contribute through our own carelessness to Hong Kong's self-immolation?

After about 1970 the main thrust of urban development shifted to the old agricultural lowlands of the New Territories. The massive decentralization of population that resulted, and the roughly simultaneous decline of farming, changed Hong Kong fundamentally. The lowlands were transformed into commuter countryside; but the uplands remained little changed, preserved for posterity in the new Country Parks.

Two photographs looking from Amah Rock (Mong Fu Shek) into Sha Tin valley epitomize these changes and indicate the speed and scale of development. The first was taken in 1970, the second in 1983.

In the first photograph, Amah Rock stands clearly above grasses and low shrubs. The head of Sha Tin valley is divided by the wide mouth of the Shing Mun River, and hills slope steeply into Tide Cove's long inlet. The lowland is patterned with paddy fields, vegetable farms, fish ponds, and villages.

In the second, Amah Rock is half-hidden under larger shrubs and trees. But the natural mouth of the Shing Mun River is gone, and a concrete channel has replaced Tide Cove. Where before a ribbon of land lined the coast, now massive reclamations are covered with new housing estates, and above the valley tower blocks march up the hillsides.

During the early 1970s Hong Kong showed a strong belief in its future. There was an increased sense of identity, as *Heung Gong yan* (Hong Kong people) developed bonds that weakened the rootless, refugee mentality. There was pride in the post-war

Government reforestation has succeeded in 'greening' Hong Kong, as these trees spreading up a Sai Kung hillside show.

improvements, and relations with China were less troubled. It was these combined factors that allowed the government to embark on visionary schemes.

This renewed optimism was symbolized by the arrival in November 1971 of a new Governor: Sir Murray, now Lord, MacLehose. Sir Murray made his first annual address to the Legislative Council a year later, on 18 October 1972. He stressed three issues of the greatest importance to Hong Kong's natural landscape: housing, Country Parks, and environmental degradation.

Since 1953 about 1,600,000 people had been rehoused by the government, but a further 600,000 squatters and others still needed proper housing. Sir Murray, to allow for the expected population growth, announced a bold ten-year plan to rehouse a further 1,800,000 people. The already densely populated urban areas could never decently house so many people: instead, satellite cities—or New Towns—would be built on the New Territories' lowlands.

Partly to compensate for this loss of countryside, Sir Murray also announced plans to create Country Parks, mostly in the undeveloped upland areas. Besides recognizing an obligation to protect the country for its intrinsic value, the government also saw a need to provide city dwellers with natural places for recreation.

Concerning the environment, the Governor stated: 'We already face serious pollution in some streams in the New Territories and in parts of the harbour, and the danger of a rapid advance in pollution is very much in all our minds.' He assured Hong Kong that, when a departmental committee then looking into pollution had reported, the government would 'take a hard look' at the need for effective controls.

The shift away from farming during the 1960s and 1970s marked a watershed in Hong Kong's agricultural development, and transformed the lowland New Territories. By 1980, less than 1 per cent of the land was growing rice, and fully 40 per cent of all agricultural land was recorded as 'abandoned'. Cheaper rice from China had hastened the collapse of local rice growing.

The farming decline, and the concurrent industrialization, had taken less than a generation. Thus rapidly were the New Territories' lowlands transformed, their patchwork of paddy fields replaced by spreading development—mostly factories and housing estates. In the process, very extensive farmland habitats were lost.

If there was a compensating gain for the countryside, it was the steadily greater reforestation of the uplands. For, with a far smaller rural population, and few people now unable to afford gas or electricity, grass- and tree-cutting diminished until they became insignificant.

One has to look no further than the plantations in many Country Parks for evidence of the post-1970 reforestation: today trees extend up all but the highest hillsides. Or one can compare earlier and later photographs of the same sites, such as two views of the Shing Mun area. In one, taken in the mid-1960s, the slopes above the reservoir have scattered trees, while the higher slopes are treeless and stark. The same place, photographed in the mid-1980s, shows dense, secondary forest over the entire valley, with trees extending far up Tai Mo Shan's flanks.

During the 1970s and 1980s Agriculture and Fisheries Department (AFD) foresters planted well over 200,000 trees most years—sometimes as many as 400,000. These were generally mixed species: Chinese Pines, Slash Pines, Brisbane Boxes, Acacias, Paper-barks, and, increasingly, native broad-leaved species. Although successful in stabilizing hill slopes and providing tree cover, ecologists now regard the earlier emphasis on exotic species as regrettable—as local insects and small fauna are not adapted to the exotic species.

None the less, two valuable gains flowed from the plantings. First, the quantity of soil eroded off hillsides fell markedly. Second, as Richard Webb has recorded, Hong Kong still has about 340 significant *fung shui* woodlands which cover in all over 700 hectares. These woodlands have gradually spread into abandoned farming areas, and propogated in nearby forestry plantations—thus helping form mature, mixed woodlands. In such woodlands there has been significant recolonization by native fauna, especially birds and terrestrial mammals. For instance, Gary Ades, Senior Conservation Officer of Kadoorie Farm and Botanic Gardens, reports an increase in mammal populations in a well-wooded area of less than 140 hectares near Tai Mo Shan. These include resident Barking Deer, Masked Palm Civets, Chinese Ferret Badgers, and Porcupines; and visiting Small Indian Civets, Javan Mongooses, and Leopard Cats.*

* *Gary Ades believes that, because of their need for 'territory', the current population of Barking Deer may be close to the maximum—given the available areas of mature woodland.*

The post-war economic growth in 'developed' countries highlighted the need to protect countryside from spreading development, especially near cities. Conservation parks came rather later to Hong Kong, partly because of its other pressing challenges in the 1950s and 1960s.

During the early 1960s a small group of academics, foresters, and naturalists, alarmed at rapid intrusions into the rural landscape, began pressing the government to act to contain the spreading 'development'. The government's first step was the commissioning of a wide-ranging study, made in 1965 by the conservationists Lee and Martha Talbot. In so densely populated a place, the Talbots argued, nature conservation must allow room for recreation. Their report therefore explained the need for a 'tiered' system of Country Parks, with areas ranging from family picnic sites to wild, virtually untouched tracts. A Commission of Enquiry following the 1967 riots also stressed the need for better recreational outlets.

The Country Parks Authority, established in 1967, made a provisional report in June 1968. Three themes dominated its conclusions. First, that recreation, nature conservation, and the reservoirs must be seen—and managed—as separate aspects of one system. Second, that urgent action was needed to halt erosion, and loss of vegetation and habitat. And third, that it was essential to arm any conservation body with effective means to fight the scourge of the Hong Kong countryside—fire.

Sir Murray MacLehose's arrival in 1971 put the Country Parks concept high on the government agenda. A keen walker, Sir Murray had known the countryside in the early 1960s—and on returning he was struck both by its new recreational use and by its degradation. E. H. 'Ted' Nichols, Director of the AFD (1965–80), shared Sir Murray's enthusiasm for the Territory's countryside. Together the two men helped ensure that the Country Parks vision became reality.

The Country Parks Ordinance of 16 August 1976 created a permanent Country Parks Board and a Country Parks Authority. Both aimed to preserve, and yet make accessible, large areas of unspoilt countryside. The former established policy, and the latter, under the management of the AFD, implemented it.

For Hong Kong, where planning had usually been incremental, the government's decision to establish the Country Parks was bold and visionary. But, given the increasing dangers from development, it came none too soon. As Sir Murray later wrote, 'without the Country Parks Hong Kong would have had no lung, no undeveloped areas, so a line had to be drawn.'

The Country Parks, by keeping popular picnic places close to their boundaries, have preserved their wilder areas, like this upland glen in Shing Mun Country Park.

To coordinate the establishment of the Country Parks, the AFD created a Conservation and Country Parks Branch. This began the map interpretation and field reconnaissance needed to 'fit' park boundaries to the topography and existing land use—an extremely difficult balancing of often conflicting interests. It drew together the research required for rounded impressions of each future park's geology, flora, fauna, forestry, and water resources. And it managed the many practical tasks: building management and visitor centres, making tracks and roads, improving communications, and widening fire-breaks.

The most vexing issue was balancing the needs of conservation and recreation with those of local villagers. Concerned at the Country Parks' intrusion on their 'domains', villagers were allowed to retain quite large areas of land—seen today as 'dog-leg' and 'horse-shoe' exclusions along the Country Park boundaries. No doubt proper at the time, these exclusions now threaten the conservation of some of Hong Kong's most beautiful areas: for, with the subsequent village depopulation and other social changes, these same areas are now prey to unprincipled, rapacious 'development'.

Twenty-one Country Parks were designated between mid-1977 and late 1979. They ranged from small parks, such as those at Kam Shan and Pok Fu Lam, to very large ones in the mainland New Territories and on Lantau. About half the Country Parks covered more than 1,000 hectares, and six more than 3,000 hectares. Numerous much smaller areas were made 'Sites of Special Scientific Interest', among them the Mai Po marshes.

The twenty-one Country Parks covered in all 40,000 hectares (400 square kilometres)—about five times the area of Hong Kong Island. This represents almost 40 per cent of the Territory's entire land area, an extremely high proportion by any standard—and doubly so given Hong Kong's chronic shortage of land. Had the government not acted when it did, much of this invaluable countryside would have been developed.

Elsewhere in the world the proportion of countryside protected as National Parks, the equivalent of Hong Kong's Country Parks, is far smaller. National Parks cover about 7 per cent of the United States, and 9 per cent of Australia. No Asian country has anywhere near Hong Kong's proportion of land vested in Country Parks: Indonesia and Thailand, each with about 7 per cent of their land so designated, have by far the highest proportion for Asia outside Hong Kong. Seen from a helicopter, the Territory's developed areas are shown to be a mere fringe around far larger areas of wild country.

While Hong Kong's uplands and more remote areas were being preserved, its old agricultural lowlands were rapidly disappearing under widespread urbanization.

The government's ten-year housing programme that began in 1973 committed it to a major redistribution of population: from the overcrowded parts of Hong Kong Island and Kowloon to the New Towns being planned for the New Territories. The New Towns were centred on the existing rural market towns—and the speed of their development was phenomenal. In 1970 future New Town areas had only about 450,000 people; a decade later they had homes for one million—and in 1990 for over two million.*

Only photographs truly reveal the speed of construction. After each New Town was begun, its lowlands were levelled or reclaimed within a few years; housing estates appeared soon afterwards; and within a few more years entire residential-commercial-industrial complexes stood where little more than a decade before the land had been agricultural. On returning to Tsuen Wan after a few year's absence, James Hayes wrote:

> *The last remaining foothills were being carved up into large platforms for yet more public housing.... The place was scarcely recognizable to anyone who had known it in the past.*

Previously other Hong Kong village lands had been taken over for the city's needs. But the earlier urban expansion into the rural landscape had been very gradual: the New Towns, by contrast, exploded into the countryside. The old villages and the new centres were starkly different, as Solomon Bard recorded:

> *After enduring for centuries, the old village life disappeared in the course of a few decades. Massive town development has fundamentally changed not only the population structure and local social traditions, but the very landscape.*

For traditional villagers of the New Territories even the construction of a house, or the cutting down of some trees, could severely affect the local *fung shui*—and so their perceived connections between the people and their environment. Now entire landscapes were being recast: hills cut down, streams filled in, bays reclaimed. The New Towns' development thus challenged the villagers' view of their surroundings. For to them, as Austin Coates wrote:

* *The initial New Towns were sited at Tsuen Wan, Tuen Mun, and Sha Tin. A 'second generation' were developed around Tai Po, Fanling, Sheung Shui, and Yuen Long; and the most recent ones are at Tin Shui Wai (Deep Bay) and Tseung Kwan O (Junk Bay).*

Large numbers of urban people now visit the Country Parks, though only the lucky few go by boat. Two decades ago bays such as this at Long Ke were rarely visited except by fisher-folk.

The New Territories' remote villages are now depopulated, if not completely abandoned. This cobwebbed temple at Cheung Uk in Sha Lo Tung reflects the changing times.

> *Every rib of every hill, every spur, every eminence ... was the abode of some sleeping dragon or tiger—was a dragon or tiger, in fact—and it would have been beyond the capacity of the most learned* fung shui *expert to have identified them all.*

The resiting of graves prior to 'development' often prompted far deeper upset than the relocation of villagers themselves. Great care was needed to ensure that new grave sites had favourable *fung shui*, and village land was only resumed following lengthy negotiations—as District Officers well knew. Village concerns about land ownership had become even more entrenched with the escalation of land values—and so compensation.

Transport systems reached into the countryside, drawing Hong Kong's regions closer together than ever before. In the past, villagers had trudged *over* the uplands rather than skirting them by longer paths. Now, engineers blasted new short cuts *under* the same slopes. Tunnels cut through the mountains and under the sea. With the opening of the Aberdeen Tunnel in 1982, and using the Lion Rock Tunnel and the Cross-Harbour Tunnel opened about a decade before, it became possible to drive from Aberdeen to the Shenzhen River without rising more than about 50 metres above sea level. Getting about became ever more convenient—ever more mole-like—and ever more cut off from Hong Kong's rugged and beautiful terrain.

Few other places can claim to have managed such a rapid, largely successful decentralization of a concentrated urban population. But the New Towns also brought clear losses to the countryside itself.

The most obvious drawback, one rarely recognized, was the loss of lowland habitats. With the end of rice growing, grasses had invaded the disused paddy fields; but the old irrigation channels still helped maintain 'wet' conditions. Thus valuable wetland habitats were retained for small fauna, among them insects, amphibians, freshwater snakes, and waterbirds. However, as the New Towns were built, these wetlands disappeared under landfill and concrete.

Far more damaging in the longer term was the insidious pollution of the countryside's land, water, and air. Even in the early 1970s, before the New Towns were built, two things were clear: the country itself produced large volumes of waste, and industrial pollution was slowly infiltrating the countryside.

Since 1946 the government's annual reports had included a perennial chapter entitled 'Geography and Climate'. This was replaced in 1972 by one entitled simply 'The Environment'. The 1972 chapter stressed the government's concern over air and water pollution, and recorded the establishment that year of the Advisory Committee on Environmental Pollution—or EPCOM. This body, Hong Kong was assured, would devise strategies to counter the escalating degradation of the environment.

One of the most critical environmental problems, and that where urban waste most directly affected the countryside, was water pollution.

The first EPCOM report (1972) had observed that 'the sea and especially the harbour [are] the city's dumping ground'. By 1976 Victoria Harbour was no longer being properly flushed of its daily dose of pollutants (partly because reclamations, despite contrary predictions, actually had weakened the tidal flow). Numerous studies documented the increasing degradation of the New Territories' coastal waters, especially Deep Bay and Tolo Harbour. Even in the mid-1970s, well before Sha Tin and Tai Po were developed, Tolo Harbour's land-locked waters were heavily polluted, severely de-oxygenated, and below accepted world standards for either fishing or swimming.*

The streams flowing into Tolo Harbour were largely to blame. Indeed, as researchers including K. C. Lam and John Hodgkiss showed, their lower courses were virtual sewers. A 1974 study of

* *The content on water pollution is mostly drawn from Dr K. C. Lam's chapter 'Environmental Problems and Management', in* A Geography of Hong Kong *(edited by T. N. Chiu and C. L. So). The government did not publish detailed data on the environment until later.*

Hong Kong consumes far more energy than before, so new power stations, such as this Lamma Island facility, are needed.

Fire was, and remains, the greatest single threat to the Country Parks. Wildfires destroy vegetation, especially shrubs and trees, and incinerate animals.

some 400 kilometres of New Territories' streams indicated that virtually all their lowland sections were either 'polluted' or 'grossly polluted'. Agricultural wastes made up about half the pollutant load, with hundreds of tonnes of livestock and poultry excreta being washed into the streams daily. Domestic sewage and industrial effluent accounted for another third of the total.

By about 1980, when the total population of Sha Tin and Tai Po was about 500,000, Tolo Harbour was a grim sight. 'Red tides' occasionally occurred, and by 1983 marine scientist Brian Morton could write: 'The inner reaches of Tolo Harbour are to all intents a dead sea'. He added more generally,

> *The real wonder, perhaps, is that Hong Kong still has so many healthy shores. The question the conservationist must ask is why they have survived, and how their continuing existence can be maintained.*

Countryside degradation also resulted from the trade with China, which expanded rapidly during the late 1970s and especially the 1980s. Storage for containers was severely limited and, especially in the north-western New Territories, developers began dumping building waste into abandoned paddy fields to form container parks. The process was judged by policy-makers to be detrimental, but the government had no effective sanctions to counter the destruction.

Town-planning and land-use zoning, which controlled urban development, did not apply in the New Territories. That the government failed to address this legal and environmental anachronism stemmed largely from an unwillingness to confront the New Territories' powerful vested interests—and the Heung Yee Kuk. Thus, lamentably, the escalating wetland destruction continued throughout the 1980s.*

During the 1970s and 1980s, Hong Kong's environmental legislation was almost all concerned with urban pollution—not rural conservation: that the two issues were inextricably linked was overlooked. Moreover, relatively little research had been done

* *By 1983 about 1,100 tonnes of construction waste was disposed of daily, making one-sixth of the Territory's daily solid waste. Paddy-field dumping minimized disposal costs.*

New reservoir roads and catchments have given hikers invaluable access to majestic landscapes such as Ma On Shan—seen here from High Island Reservoir, completed in 1979.

into local ecology. Thus there was only scarce ecological data to help monitor the impact of rapid development.

Despite the growing evidence of degradation, too little was done to halt—let alone reverse—the deterioration of the natural landscape. 'Action to improve the quality of the environment inevitably involves expense', an independent 1975 report had warned. Yet throughout the later 1970s and 1980s, though Hong Kong incomes and wealth grew apace, only very meagre amounts were spent to safeguard the Territory's countryside and urban environments. The government, so effective in solving immediate crises, lacked the will to stem longer-term environmental degradation.

But the government was only partly at fault. Indeed, its concern was well ahead of community attitudes, and its initiatives were often hampered by public apathy and industrial opposition to disturbing environmental reports. 'Bugs here, there and everywhere—Hong Kong to Drink Raw Sewage?' one newspaper headline loudly challenged. But the Territory's *laissez-faire* economic growth, and sheer complacent greed, remained largely unchanged until the late 1980s. By then the damage was done.

Some academic environmental research reached a wide audience, such as Hong Kong's first ecology textbook, published in 1981 and written by John Hodgkiss, Stella Thrower, and S. H. Ma. But large volumes of research remained hidden in academic and specialist reports. By contrast, general accounts of the Hong Kong countryside became widely available during the 1970s and 1980s. The Urban Council, reflecting increased interest in the Country Parks, published numerous informative booklets describing the local flora and fauna.

One of the 1970s most striking countryside publications was *Wild Flowers of Hong Kong* by Beryl Walden and Hu Shiu-ying. This presented beautiful, botanically accurate water-colours of the local wild flowers. One was MacLehose's Cymbidium, a rare mountain orchid that flowers in the early winter with long creamy petals—named after Lady MacLehose in 1974.

Sir Murray and Lady MacLehose both knew the local countryside well. In her Foreword to *Wild Flowers of Hong Kong*, Lady Noel MacLehose wrote:

> *When I was in Hong Kong from 1959 to 1962 and used to walk on the hills of the New Territories one rarely met anyone; and those one did meet were all country people—grass-cutters, cow-herds or people working in the high fields—scarcely ever people from the city.*
>
> *When I returned to Hong Kong in 1971 I found that the cow-herds and grass-cutters had gone, and many of the villages they lived in had been deserted. But on the hills, and even on the most inaccessible ridges and peaks at weekends, there were now young people from urban Hong Kong.*

Increased education and affluence in the 1970s and 1980s led to a steady growth in countryside recreation. Well over two million people visited the Country Parks in 1977, when the first parks were gazetted. Two years later the figure was over five million; in 1982 and 1983 it reached seven million; and by 1988 to 1989 it was over nine million. The more distant Country Parks were the most popular; but the overwhelming majority of visitors stayed near accessible picnic areas—and went no further.

The Country Parks remained under the umbrella of the AFD; and, though frugally funded, their access and facilities were steadily improved. As initially planned, the Country Parks offered 'zones' that allowed people to enjoy either family picnicking, moderate hiking, or true 'wilderness' experiences.

That Hong Kong has almost 40 per cent of its land vested in Country Parks is misleading; for, given its small total area and high population, the amount of Country Park per capita is very small—only about 70 square metres per person. Indeed, for the best 'natural' country to be conserved, it was in fact essential that the vast majority of visitors be 'channelled' into limited (easily serviced) 'picnic' areas—a policy which happily coincided with the desires of most Hong Kongers.*

For the more adventurous, the MacLehose Trail showed how much rugged country Hong Kong still had. 'The MacLehose', a 100-kilometre hiking trail that crosses some of the highest and most spectacular country, was the brainchild of the then Director of 'Ag and Fish', Ted Nichols. As Sir Murray MacLehose recalled in his Foreword to *The MacLehose Trail*, Nichols had the old New Territories paths marked and mended: 'They eventually joined up the village and grass-cutters' tracks to make a continuous path through open countryside from Mirs Bay to the Pearl River'. 'Trailwalker', an endurance hike along the entire MacLehose Trail, which began in 1982 as a military exercise, in 1986 became an annual fund-raising event—and is now a Hong Kong institution, consistently won by tough Gurkhas.

Excluding fire, erosion and littering were always the Country Parks' greatest problems. Hiking can easily weaken erosion-prone hills, which then require costly stabilization. Combating littering, through education and cleaning up, is simpler—but costly and

* *Much of this information comes from the urban ecologist Dr C. Y. Jim's valuable paper, 'The Country Parks Programme and Countryside Conservation in Hong Kong'. The paper was published in the* Environmentalist, *Vol. 6, No. 4 (1986).*

time-consuming. In the winter of 1978 to 1979, for example, over 2,000 tonnes of litter were removed by Country Parks workers; five years later the amount had almost doubled. Littering did not generally damage the flora and fauna. However, as Stella Thrower wrote: 'The illusion of "being at one with nature" ... is rudely shattered by empty bottles'. Littering remains a seemingly intractable problem.

Fire was always—and still is—the Country Parks' greatest menace. Throughout the 1980s over 10 per cent of their total area was burnt during the driest winters, and about half that during average ones. The killing of trees and shrubs, the incineration of small fauna, the destruction of habitats and food supplies, and increased erosion were the end results.

Most hillside fires now occurred at weekends—and picnickers and people visiting graves accounted for most (lightning almost never ignites fires in Hong Kong, as it accompanies summer rain). The Country Parks tried to prevent wildfires by planting fire-resistant species, reducing undergrowth, cutting fire-breaks, and doing controlled burning. Yet with depressing regularity, fires continued to wreak destruction.

In some years more trees were burnt than government foresters could plant out—and between 1970 and 1985 the total area burnt was twice as large as the entire area of woodland existing in 1985. As David Dudgeon and Richard Corlett comment: 'The general public does not see hill fires as the disaster they are, precisely because they are so common'.

Hong Kong's climate remained virtually unchanged since a century ago—indeed, since prehistoric times. Between 1961 and 1990, the mean annual temperature was 22.8°C, ranging between January's mean of 15.8°C and July's of 28.8°C. The mean annual rainfall was 2,214 mm, almost 80 per cent of it falling in downpours from May to September. Relative humidity showed a similar seasonal pattern.

However, distinct microclimates had developed amid the urban canyons. Concrete surfaces altered the radiation; tall structures modified winds; buildings gave off warm exhaust gases, and pollution haze trapped the heat. Thus average night-time temperatures around the Royal Observatory, in the heart of Kowloon, increased by 0.36°C on average during the decades between 1946 and 1980—the period of Kowloon's most rapid urbanization.

Yet even in Kowloon the summers were less trying than before, for by the 1980s air-conditioning was almost universal. Sultry, sleepless nights gave way to heavy electricity consumption—and high bills!

Indeed, by the late 1980s Hong Kongers had become so urbanized that most seemed hardly aware of the natural world around them. During the summers air-conditioners cooled (or chilled) people; covered walkways kept city-folk dry during daytime downpours; and on sultry evenings windows closed for air-conditioners kept flying termites outside.

For most people only the occasional intrusion of water shortages, floods, and typhoons reminded them of nature's arbitrary power.

There was no major water crisis such as that of the early 1960s during the years 1970 to 1990, because of the much greater reservoir capacity, kinder seasons, and water supplied from China. The press, however, still kept a wary eye on reservoir levels. Good rains led to cheerful headlines, such as 'No Worries Water Stores Boosted'—and dry periods brought forth warnings such as 'Water Chief Warns Against Optimism'.

Typhoon seasons brought forth countless articles. Each summer Royal Observatory meteorologists explained their maps yet again, for those perpetually mystified by the swirling concentric lines that indicate a developing typhoon.

Novel insights were sometimes given into the destructive power of typhoons. The *South China Morning Post*, perhaps not very helpfully, reported in June 1977: 'One "fully developed" typhoon has the energy of 400 20-megaton atomic bombs dropped in one day'. More traditionally, a local *fung shui* expert predicted a summer of 'death and destruction' in 1973:

> *I don't want to frighten anybody, but there will be lots of bad typhoons this year. The reason is that the year of the Ox in the Chinese calendar coincides with the symbol for water. Water means rainfall and rainfall, I'm afraid, means typhoons.*

Now typhoons became invested with human qualities: 'Fickle Iris Flees'—'Killer Kate Heading Here'—'Cora Looses Her Eye'—'Ships in Distress in Norah's Wake'—'Weak Winona Moves West'—'Flossie's Freighter Victim Freed'. By the 1980s sexual equality had led to alternate typhoons being given male names: 'Killer Percy Moves Closer'—'Gerald Growls Back'—and, happily, 'Roy Loses Strength'.

By the 1980s Hong Kong was justifiably proud of its achievements. Yet despite its urban record, notably housing the post-war population, and despite its rural record, notably establishing the Country Parks—despite those and many other advances, its people retained a startling indifference to their environment. A population without deep local roots, a culture of family and individual self-interest, and colonial attitudes that did little to encourage widespread community and personal responsibility, contributed to this apathy. Powerful business groups mostly acquiesced in the environmental complacency, and failed to press the government to take resolute action.

Various 'green' groups articulated Hong Kong's environmental woes during the later 1980s. For too long, they argued, Hong Kong had tolerated an economy that put its faith in 'disposal'—an economy that, despite its meteoric growth, continued merely to 'transfer' wastes away from their source. As the geographer K. C. Lam wrote as early as 1983:

> *The story of Hong Kong has been one of economic success, which is attributed largely to its* laissez-faire *policy. Yet, in managing Hong Kong's environment, there is no place for such a policy. Nor can there be any hope of success should the present technocratic, short-term, palliative approach to pollution control continue. The mounting problems in Hong Kong demand a drastic and comprehensive long-term environmental policy.*

The need for urgent environmental action was pressed home by Hong Kong's 'green' groups. These became increasingly prominent during the later 1980s, reflecting the growing local interest in participatory politics.

During the governorship of Sir Edward Youde (1982–6) negotiations concerning the expiry of the New Territories' lease in 1997 dominated Hong Kong. With uncertainty about the future lessened by the 1984 Sino-British Joint Declaration, a serious start was at last made to reverse the environmental decline. In 1986, with the formation of the Environmental Protection Department (EPD), the government brought together and upgraded its environmental decision-making—thus promoting long-term strategies to prevent, not merely clean up, pollution. One of the EPD's first steps was to begin introducing Water Control Zones, and in 1987 Tolo Harbour was the first.*

Sir David, now Lord, Wilson is widely regarded as Hong Kong's first 'green' governor. He was appointed in April 1987, following the tragic death of Sir Edward, and, like Sir Edward, he was a keen countryside advocate. On 11 October 1989 Sir David delivered a landmark address to the Legislative Council. The speech, entitled 'A Vision of the Future' documented developments that in the following decade would radically affect Hong Kong's natural surroundings.

* *The Environmental Protection Department grew out of a previous, smaller body, the Environmental Protection Agency, which had been formed in 1981.*

Sir David, referring to the recent White Paper on pollution, 'A Time to Act', outlined a bold ten-year programme to restore the environment. He reported some modest gains since 1986, while warning that much more remained to be done.

Sir David also announced the exceedingly ambitious Port and Airport Development Scheme—to be carried out on Lantau Island. Any project of such daunting complexity and scale inevitably would have the most profound impact on the surrounding sea and land—on their flora and fauna—and on Lantau's scenic grandeur.

Constant change had become synonymous with Hong Kong. Indeed, it was now the norm. As one longtime resident said in 1988: 'If you cannot live with change Hong Kong is not the place to live at all.' Nowhere was this truer than in the Territory's natural landscape. There, change too often had not been for the better, as Richard Irving and Brian Morton observed in 1988:

> *Though Hong Kong is a relative newcomer to the problems of environmental degradation and habitat destruction through pollution, it is also clear that lessons already learned elsewhere have only slowly been recognized by a community more intent on rapid development.*

By the 1980s polluted streams were a common sight around Hong Kong. Many were far more severely degraded than this.

Twilight on the Wang Leng hills, looking northwards over Double Haven, and across Mirs Bay to Guangdong. Descending from here in the dark I lost my way.

REGION

THE REMOTE NORTH-EAST

I gaze at the peaks, confused by their huge image,
I look at the crags, bewildered by their steepness.
Their power is so great that it hides the heavens,
They rise on high, soaring up wall after wall...

CHANG YEH, about AD 400

Night has fallen and I have lost my way. Not seriously: Hong Kong is too small for that. None the less, I am 'bushed'. Unable to locate my rucksack I face the prospect of a night without food. And, unless I can find it, I cannot climb a peak a few kilometres away for tomorrow's dawn.

After sunset I had lingered on the crest of the Wang Leng range above Plover Cove to photograph the coast by twilight. Then I hiked down into a nearby valley, heading towards where I had earlier left my rucksack near a stream. Descending through the gathering darkness, I followed the same track I had come up on.

But down in the valley the track led into a gully overhung with trees. I could hear the stream near where I had left my gear, but thick vegetation completely blocked my way. Had I mistakenly come down the wrong track? Twice I tried to find a path through, and twice I failed. Now, slightly spooked by glow-worms flashing in the blackness, getting edgy and hasty, I decided to climb uphill again and use the light from a crescent moon to try to get my bearings.

Up on the hill, the puzzle slowly resolves. I *am* on the right track, as the features around me—two peaks and a gorge—are orientated as before. So somewhere further down there must be a side-track leading off the main one, which I missed in the dark. I eat my last orange then walk down, searching by torchlight for a tell-tale break in the shrubs and trees. Finally I find a gap and push through—no?—try a bit further—yes, that oddly shaped branch—that sharp turn—*yes*, those boulders leading down to the stream—my rucksack!

North-east of Tai Po is some of Hong Kong's wildest country. Running west-to-east, Pat Sin Leng commands the landscape, the range's rugged escarpment rearing over the lowlands. Further east, a steep-sided peninsula runs eastwards past Plover Cove Reservoir to the islands that form Double Haven.

Centuries ago these remote north-eastern coasts supported a pearl fishery. Much later Hakka clans settled in the region, but it was never more than thinly populated. Even today most of the uplands here are still wild and empty. Pat Sin Leng Country Park covers over 3,100 hectares, and Plover Cove Country Park (with its extension) over 5,000 hectares. Very few people live there, although the area is larger than the whole of Hong Kong Island. The Wilson Trail, a new south-to-north hiking trail, runs through Pat Sin Leng Country Park.

A little to the east of Tai Po a low pass leads into Sha Lo Tung valley, from where a track ascends to the crest of Pat Sin Leng. Further east the range drops steeply to some streams at Bride's Pool, where one crosses the only public road. From there a wide peninsula stretches eastwards for about ten kilometres, past Plover Cove Reservoir, along the Wang Leng hills, and so on to Double Haven and Mirs Bay.

Grassland and shrubland dominate Pat Sin Leng Country Park, the result of tree-cutting over the centuries. Across much of the high country the shrubs are only stunted—and sometimes absent altogether. The only areas of woodland are in the lower gullies and valleys, though reforestation is spreading slowly in some higher areas.

Far more striking than the flora is the terrain. Pat Sin Leng (Eight Fairies Range) runs west-to-east for five kilometres, its ridges averaging 600 metres and rising to 639 metres at Wong Leng. The range is wild and grand. To the north it steps down slowly to Starling Inlet (Sha Tau Kok Hoi), but to the south a cliffy escarpment plummets past rubble-strewn slopes to Sha Lo Tung

Grand mountains and dramatic skies typify Pat Sin Leng, the jagged range that rears up north of Tai Po.

valley. The north-flowing streams descend slowly, but the south-flowing ones cascade through short, steep ravines.

Pat Sin Leng, everywhere dramatic and in places dangerous, is formed from volcanic and sedimentary rocks. The southern side of the range is entirely volcanic, and has been severely eroded—so creating its sheer escarpment. The northern side, however, has conglomerate lying over volcanic material. The conglomerate, tough sedimentary rock embedded with stones, has resisted all but the slowest erosion—and so retains its relatively gentle slopes.

Mountain ranges, generally, are what remain after their surrounding rock has been eroded away to form valleys: that much is comprehensible. But peering gingerly off Pat Sin Leng's escarpment, gazing down hundreds of metres into the Sha Lo Tung valley, the process of mountain and valley formation is scarcely believable. Realities blur. The best afternoon light is now two hours away: where should I try to photograph this landscape? Sha Lo Tung was formed countless millions of years ago: how can one conceivably capture so vast a time span?

In Sha Lo Tung (Sand Shell Cave) the larger fauna have gone long since, but its valley remains an ecological storehouse. However, with species often concentrated in small areas, individual ones can easily be lost.

In this century and well before, tigers and leopards were probably never more than visitors to Hong Kong—though once they were native. Yet, despite centuries of agriculture, they were still sometimes seen in the early 1900s around Sha Lo Tung.

'Madam stripes and her cubs [may appear] at any time during the autumn', R. C. Hurley noted early this century. Tigers still occasionally appeared in the New Territories' wild north-east in the 1930s, having roamed in from Guangdong—usually during the winter. Leopards were much rarer but, as G. A. C. Herklots recorded, one was captured near Pat Sin Leng on 20 December 1931. Local villagers, their traps set for deer, ensnared a leopard instead. It was shot and dismembered:

> *The beast was skinned and the flesh, bones, skull, teeth, whiskers and claws were sold to bidders amongst the other villagers for the large sum of $150.*

Sheer slopes sweep down from the Pat Sin escarpment into Sha Lo Tung. In the distance, old hillside terraces are slowly being lost under spreading vegetation.

Many smaller animals continued to flourish amid the region's deforested agricultural landscapes. The paddy fields, in particular, provided valuable wetland habitats—for insects, frogs, worms, small fish, freshwater snakes, and birds. The patchwork paddy fields supported countless waterbirds, and across the New Territories there were significant egret marshes. But, like the forests before them, the rice paddy habitats now have gone.

This loss of wetland habitats in recent decades had a drastic ecological impact, especially on the populations of waders. Besides the Mai Po area, the only remaining egretry in Hong Kong is now at Yim Tso Ha, north of Pat Sin Leng. The birds there are protected and appear to be increasing. Among them are Chinese Pond Herons, Night Herons, Great Egrets, Little Egrets, Cattle Egrets, and extremely rare Swinhoe's Egrets.

Two decades ago the older villagers were still farming around Sha Lo Tung. But the younger generation—lured by urban wages and comforts—were abandoning the valley. Sha Lo Tung's three villages today testify to the migration that since about 1970 has emptied almost all the New Territories' more remote areas. Cheung Uk, the largest Sha Lo Tung village, now has only a handful of permanent residents; Lei Uk is empty though still standing; and Ping Shan Chai is decaying and overgrown.

The track to Cheung Uk begins near the Tai Po Industrial Estate, and continues uphill past unsightly container parks and car wreckers. Backward glances show factories and housing estates set amid an ever-changing landscape of levelled hills and reclamation. Then one is beyond the pass, striding down into peaceful Sha Lo Tung, along to Cheung Uk.

The village is set below its hillside *fung shui* wood. There are cobwebbed houses, there is a temple. Through chinks in bolted doors the past is glimpsed: beds, bowls, chopsticks, hats, photographs.... Possessions left behind to appease the spirits? Or left because taking them would drag the past into the future? Flats have washing machines not enamel bowls, flat dwellers wear Walkmans not rattan hats.

Pat Sin Leng is one of the Territory's most rugged ranges. Its peaks sweep down to deep valleys and command airy views.

Tiny Hok Tau Reservoir, at the northern end of the Sha Lo Tung valley, is seen here from halfway up to the Pat Sin escarpment.

Grasses and shrubs cover the uplands of Pat Sin Leng Country Park. In its higher parts grasses dominate over scattered shrubs.

Cheung Uk's half dozen old people eke out the days and weeks: poking around the village, pulling up invading grasses, waiting for decrepitude to descend on the village and on themselves—remembering. At Cheung Uk I always wish I had time to stay, and sufficient language to listen to the villagers' stories. Oliver Goldsmith's *The Deserted Village*, with its 'widowed, solitary thing', always comes to mind:

She only left of all the harmless train,
The sad historian of the pensive plain.

Anywhere in the world, Sha Lo Tung would be striking. In crowded, ever-changing Hong Kong the valley is remarkable—and beyond price. Ranges enclose Sha Lo Tung valley on all sides, shutting out the New Towns only kilometres away. The villages are surrounded by overgrown land—paddy fields that only a few decades ago held water and rice.

Gazing down into Sha Lo Tung, reflecting on its ecological life and human history, it seems impossible that the valley might not be preserved as it is—that this reminder of Hong Kong's natural and agricultural heritage might be squandered for 'development'.

Yet some years ago Friends of the Earth was forced to mount a court challenge to a *government* planning board's decision to allow a golf course resort to be developed in Sha Lo Tung, in an area originally not included in Pat Sin Leng Country Park. The development, first mooted much longer ago, would severely impact on the natural landscape—and destroy much of its rustic appeal and ecological value.

Resulting studies revealed the richness of Sha Lo Tung's fauna populations, and showed that grasses growing on the old paddy fields attract large numbers of insects, birds, and bats. The dragonfly and damselfly populations are striking, including over half of Hong Kong's 100-odd species.

As a result, the development proposal was altered to a residential complex—but even that would be an intrusion. Then, with the matter still undecided, in June 1995 workers hired by the local villagers began bulldozing six hectares of valley land. The villagers claimed, amid patent subterfuge, that they planned to return to subsistence farming!

A government spokesman admitted that, while the *construction of buildings* on private land required official approval, the *flattening of the land* did not. The issue remains unresolved.

One must wonder, perhaps one must ask: did the project developers—or the government planners—ever climb up to Pat Sin Leng and look out over Sha Lo Tung? Or did they simply grasp at some 'empty' lowland marked on the maps? Did the villagers—mostly absentee land owners— pause to consider anything except their immediate financial gain?

Like other proposals for pockets of land bordering the Country Parks, this Sha Lo Tung development would benefit only a handful of people. Other 'low-impact' uses of the valley, such as a Museum of Farming History or an Ecological Field Centre, would benefit far greater numbers.

The creeping rise in sea levels after the last Ice Age created Tolo Channel and Plover Cove. Indeed, nothing else could have produced the strange double-ended peninsula that encloses Plover Cove and runs out to Bluff Head. This finger of land, barely a kilometre wide and often much less, is some ten kilometres long.

Plover Cove (named after HMS *Plover*, a survey vessel) was drawn into the reach of urban Hong Kong in the 1960s. The inlet's sheer size, and the streams flowing off Pat Sin Leng and Wang Leng into it, had the potential for a reservoir larger than any existing Hong Kong water storage.

It was while swimming in Plover Cove itself that the concept behind the reservoir reputedly came to T. O. Morgan. Then in charge of Hong Kong's water supplies, Morgan saw beyond conventional topography and hydrology to conceive a project of breath-taking simplicity yet daunting complexity. By enclosing Plover Cove behind two dams, the inlet could be turned into a reservoir—without the loss of land that flooding a valley demanded.

Work began in the early 1960s. Villagers then living near Plover Cove were relocated, their homes lost to provide water for the entire population. Two photographs, both published in the 1965 government report, hint at the clash of interests.

In one, three Plover Cove villagers stand before a shrine. The men, in baggy peasant clothes, old black umbrellas furled, are making offerings of oranges and joss. Preparing to be moved from their ancestral village, Sam Mun Tsai, they worship for the last time beneath hills of midsummer green—beneath hills permeated with *fung shui*.

The subject of the other photograph is the *Biarritz*, a dredging barge with, we are told, 'the world's largest grabs'. A slurry of mud is draining from one of them, as the dredger excavates Plover Cove's seabed prior to 'bedding' one of the reservoir's two dams.

By early 1967 Plover Cove's two dams were complete. The salt water they enclosed was pumped out and the seabed flushed. By October that year fresh water was flowing into the maze of pipes and tunnels linked to the urban areas. The reservoir's initial capacity was 180 million cubic metres, but in 1973 its dams were

Streams rising in the eastern Pat Sin hills mostly flow into Plover Cove Reservoir, after cascading through Bride's Pool—seen here after summer rains.

A dank, moss-covered cutting near the tiny Hok Tau Reservoir, close to Cheung Uk village.

raised to increase the capacity to 230 million cubic metres—over nine times that of Shek Pik Reservoir on Lantau.

Plover Cove Reservoir covers a greater area than the whole of Kowloon. Viewed from the dam at Tai Mei Tuk its size is impossible to grasp. But climb the 300-odd metres up to the Wang Leng escarpment, right above the reservoir, and its entire, majestic expanse lies before you.

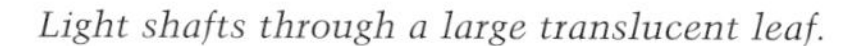

Plover Cove Reservoir, despite forcing the relocation of villagers and modifying Tolo Harbour's tidal flow, was an environmental gain: it provided essential water and also scenic beauty. The same cannot be said for subsequent developments affecting the waterways from Tolo Channel to Double Haven. Many centuries ago these same waters were rich in pearls, and decades ago they were still translucent: 'green shallows' with 'fringes of white foam' as Graham Heywood saw them in the 1930s. That is no longer the case.

From the Wang Leng peaks, to the south one looks into Tolo Harbour and along most of Tolo Channel or Chek Mun. Across the channel Three Fathoms Cove or Kei Ling Ha Hoi nestles below steep hills; but Jones Cove or Pak Sha O, site of the new Hoi Ha Marine Reserve, is hidden. To the north Double Haven or Yan Chau Tong is enclosed by encircling islands, which mark the edge of Mirs Bay or Tai Pang Wan.

Seen from Wang Leng, this coastal panorama is of startling intricacy and beauty: the blue-green waters seem timeless, changing only with the tides and weather. But the reality is otherwise: for, to varying degrees, these north-eastern waterways are all now polluted.

The despoliation began long ago. From about 900, for some four centuries, these north-eastern waters were harvested for pearls—then so abundant that the coast became known as 'The Pearl Pool'. Various Chinese governments controlled the pearl gathering, whose fluctuating harvests reflected—more than anything else—the wisdom or greed of the current emperor. By early in the Ming Dynasty excessive exploitation had virtually wiped out the pearl oysters, and pearl diving ended.

This clear environmental loss was the direct result of human folly. But the process had taken centuries. Today's degradation of the same waters has taken mere decades.

The mechanization of the fishing fleet in the 1960s led to serious over-fishing throughout Hong Kong, especially inshore. The roughly simultaneous agricultural shift to pig and poultry rearing from rice growing produced, rather than absorbed, manure—which was washed into streams, and so into Tolo Harbour. The pollution escalated alarmingly with the completion of Sha Tin and Tai Po New Towns in the 1980s, for their huge volumes of domestic and industrial effluent flow, after minimal treatment, into Tolo Harbour. The tides, ebbing and flowing through Tolo Channel, distribute the waste.

Light shafts through a large translucent leaf.

A superb midsummer dawn on Tiu Tang Lung. Beyond Double Haven, the waters of Mirs Bay extend for 15 kilometres to the coast of Guangdong.

The same morning, soon after dawn on Tiu Tang Lung.

The summit of Tiu Tang Lung looks down on Hong Kong's old agricultural lands and the village of Kop Tong. Further afield are the Territory's modern trade routes into China, where development is consuming the distant Guangdong coast.

Hong Kong's north-eastern waters have always had particularly varied marine habitats, and so have been rich in species. This was the region with true ocean water, clear and salty—yet with sufficient coastal streams to create less saline inshore niches. And, as Brian Morton says, as elsewhere in Hong Kong there was 'a whole spectrum of seashore habitats that, at first glance, defy definition': exposed, rocky coasts; sheltered, sandy shores; tidal mangrove flats; and even corals.

Along the South China coast, for purely ecological reasons, corals generally exist only around the offshore ocean islands. Yet, until recently, Hong Kong had some fifty coral species, and twenty-six coral genera. The greatest number of these species, and the most abundant corals, were in the north-eastern waters.

Corals can only grow and survive in waters that conform to specific conditions of depth, light, salinity, temperature, and 'turbidity', or the amount of suspended material (whether silt or pollutant particles) in the water. Hence corals, unable to move elsewhere like fish, provide a mirror—and sometimes a measure—of changing marine conditions.

Tolo Channel was fringed with corals until the early 1970s. Indeed, when Plover Cove was drained in 1967, numerous corals were exposed. But as the pollution levels have increased there has been a steady decline both in coral species and numbers in the channel. As Brian Morton and Joanna Ruxton write in *Hoi Ha Wan*, there is now a dramatic contrast between Mirs Bay's relatively unpolluted water and Tolo Channel's severely polluted water. Around Mirs Bay, and at Hoi Ha Wan, there are still up to thirty-six coral species—and most of them are often seen. But along the inner Tolo Channel there are now only one or two species, and even these are rarely seen.

How could it be otherwise? Years ago, about 1960, I sometimes canoed around Plover Cove: on good days, close inshore, you could see right down to the seabed. But by the early 1970s, a decade before the New Towns were developed, one study of Tolo Harbour near Sha Tin found the seabed 'covered with anaerobic muds with few living organisms'. And now the waste from over 100,000 people enters Tolo Harbour every day.

Tiu Tang Lung (Hanging Lantern), 416 metres, looks directly over Double Haven. I had climbed the peak twice before, but both times the hazy, polluted atmosphere had veiled Double Haven's grandeur. So, after a few half-hearted photographs, I had given up. Instead I had visualized the image I wanted (and the clean Hong Kong I wanted?). This ideal photograph had sharp light etching Double Haven's ring of islands—against water that gleamed with a pearl-like lustre.

Months later, when days of summer rain had washed the pollution 'particulates' from the air, the atmosphere was crystal clear. Now was the time to photograph Double Haven! But on my way to Tiu Tang Lung I became lost, as the opening of this chapter described.

It is ten o'clock that same night. I have found my rucksack, but the delay has cost two hours—and I cannot now face climbing Tiu Tang Lung. But, if I reach its base tonight, I can climb to the summit before dawn. The sky is clear, so I set off.

There is just enough moonlight, with occasional jabs of torchlight, to find my way. I cross a stream where cattle are drinking, then climb up a rocky shoulder to another path. This one, I know, is narrow and the slope drops steeply. A track leads off to an abandoned hamlet, a Hakka place. My rucksack and camera gear seem lighter as I think of the Hakka: swaying under heavy loads, heading home through these same hills on other still summer nights.

It is almost eleven when I reach the spur below Tiu Tang Lung. I am exhausted. Barely bothering to smooth the ground, I bed down in a hollow. The summit is 250 metres higher, probably about an hour's climbing in the darkness. Gazing up at the stars, I half wonder if this book has been worth the effort. An unexpected answer echoes across the valley: a faint bark ... another ... and then silence. This time I know the sound: a Barking Deer—50 kilometres from the one I heard at another extremity of Hong Kong, Fan Lau on Lantau.

My alarm buzzes at a quarter to four. The moon has set; it is pitch dark but the sky is clear. Climbing Tiu Tang Lung demands care and a clear head—so I do some physical jerks, throw water over my head, eat a sandwich. Now: go slowly!

I had forgotten quite how precipitous and rough the track was. Ten minutes up: I am heaving for breath and sweating heavily. Another ten minutes: the sheer slopes around me are beginning to take form. My ankles occasionally give way—can I make the summit by dawn? Higher still, I see a rope snaking up out of sight. Strange: I tug it and the rope holds. I tug harder—it still holds! Since I was last here people—hikers? climbers? Gurkhas?—have put in a rope. Pulling myself up, hand over hand, I climb twice as fast and my fear of falling fades. Keep moving ... steadily ... the last 50 metres, just below the summit now.... A survey marker looms up: Tiu Tang Lung, 416 metres. It is 5.05 a.m.

I slump down, barely aware of the panorama. The sun will rise in half an hour, but pre-dawn colours are showing already. Within minutes subtle hues begin to pattern the sea: greys and mauves and pinks. Shapes begin to resolve clearly: coves, bays, islets, islands, promontories. Across Mirs Bay majestic clouds frame the coast of China. The visibility is crystal clear, and the waters take on a rich sheen, like pearl.

Tiu Tang Lung, like many Hong Kong peaks, is a place of extreme contrasts. Its summit looks down on natural grandeur—and on degradation.

It is now mid-morning. The wind barely ruffles Double Haven, sheltered by its encircling islands. Dwarf mangroves mark the heads of muddy inlets where steep hills fall to the water. Immediately below the summit, nestled in a deep valley, is the village of Kop Tong (Frog Pool). Bright green patches indicate its abandoned paddy fields, shadows trace its old hillside terraces.

Looked at differently the view changes. To the south, tower blocks almost surround Tolo Harbour: from below Ma On Shan, past Sha Tin, round to Tai Po, and on to Fanling just visible over a pass. From the border at Sha Tau Kok, stretching out around Mirs Bay is a long line of development. Dynamite blasts constantly rumble across from Guangdong. Binoculars show hillsides being torn down one by one.

North of Pat Sin Leng, the road from the Sha Tau Kok border crossing to Fanling is now mired with container dumps. The crossing opened in 1985, and is now a major truck route. But, without space for their containers, transport operators invaded the old paddy field lowlands, and turned the countryside into an unsightly mess. The government, seemingly impotent, watched as the lowlands were desecrated.

Tiu Tang Lung encapsulates the choices now facing Hong Kong. Can the Territory's remaining wild country be preserved—or will it be sacrificed to never-ending 'development' and 'growth'?

By the 1900s, urban developments dominated the New Territories' lowlands leaving only pockets of agricultural land. Here, Sheung Shui New Town towers over some remaining market gardens and fish ponds.

HISTORY 1990 and Beyond

VANISHING WILD PLACES

The present generation has the right to enjoy the countryside, but also the moral obligation to keep it as an undiminished resource for posterity.

C. Y. JIM, 1986

Hong Kong retains a truly remarkable array of animals and plants for such a small land area. Yet Government still lacks a comprehensive conservation policy.

DAVID MELVILLE, 1993

We are urban animals who have lost touch with nature. We cannot see that our own lifestyle and consumer habits have an impact on nature, depriving it of its own sustainability.

MEI NG, 1995

The 1990s began in Hong Kong with unsurprising weather. As usual, a cold dry winter faded to warmer damper months, which gave way to the hot humid summer. August 1990, however, brought exceptional conditions. The month was the hottest and driest August on record, and the 36.1° Celsius recorded on 18 August was the Royal Observatory's highest ever temperature.

Unusually dry weather continued into 1991. The reservoir levels fell, and by midsummer water restrictions were being considered. When a tropical storm faded away that July without giving any helpful rainfall the *South China Morning Post* commented: 'Hong Kong was left with only damp feet, facing the prospect that within a very short space of time its headlong, profligate lifestyle was going to be tempered a little. Water restrictions will probably come into force some time next month.'

Only those of middle-age or older clearly remembered the harsh water rationing of the early 1960s, now almost a generation ago. By August 1991 the reservoirs held just one-third of their usual summer volume; saunas and massage parlours began installing storage tanks; and the sale of plastic buckets, virtually unknown items in the 1960s, boomed. The potential crisis was averted when in late August China agreed to double its water supply to Hong Kong—and when unusually heavy rain fell that October.

Three years later, in 1994, the problem was not too little rain but too much. Incessant downpours in July brought almost four times the usual rainfall making that month the wettest July since 1884. During one twenty-four hour period the Tai Mo Shan weather station measured an amazing 875 mm of rain. The Hong Kong lowlands were inundated, while in Guangdong severe flooding brought devastation, hunger, and loss of life.

Today, Hong Kong has dense urban centres and wild country—but few farming areas. The Territory's rugged, empty uplands drop down to lowlands now largely covered with buildings. This central fact of the contemporary landscape is its unique characteristic.

Hong Kong's lack of usable land always imposed severe challenges. However, by the 1990s, its compactness also held out potential for 'green' lifestyles, with short travelling distances and easy access to countryside.

During the early 1990s some progress was made in improving the environment, but ongoing pollution made a rapid advance impossible. By 1990 the population was about 5,800,000. This large, mostly affluent population produced massive amounts of waste for so small an area, and extensive development placed extreme pressure on the countryside. Meanwhile, economic growth had transformed neighbouring Guangdong Province.

Two central issues hang over the future of the Hong Kong countryside: conservation education, and the priority which local people give to their environment; and the post-1997 administration's attitude to and controls over 'development'.

By the 1990s agriculture was virtually a lost lifestyle in Hong Kong. Pockets of farming land remain, but the only commonly seen water buffalo are the bronze animals standing near Exchange Square, in the shadow of the Stock Exchange. In the New

Much of the New Territories' lowlands are now marred by unsightly 'sprawl'—with truck depots, container dumps, and ramshackle light industry.

Territories, the old agricultural lowlands are almost all intensely urbanized or covered with unsightly sprawl.

More old paddy fields were degraded during the early 1990s. Indeed, the filling-in of these wetland habitats dominated the north-western—and increasingly the north-eastern—New Territories. Villagers desecrated their ancestral rural lands to make quick profits, and container dumps became the stark symbol of the lowlands.

Having lost sight of traditional Chinese precepts of living in harmony with nature, and having abandoned the *fung shui* dictates that guided their parents, countless villagers set out to rape their own lands. They, and the powerful Heung Yee Kuk, rounded angrily on those who dared challenge their activities. Environmentalists and government officials alike were damned, as one meeting reported in the *South China Morning Post* indicated:

> *The villagers well knew the vast cash harvest they could reap if allowed to despoil their ancestral fields.... Enraged by suggestions that this was not good for the environment they ordered [the government] town planners out of the hall.*

The government, to its lasting shame, took no positive action for over a decade. Finally, in 1991, it sought to counter this lowland degradation by amending the Town Planning Ordinance. However, only container dumps built after 1991 fell within the law—and weak enforcement led to warnings almost always coming after the event. By then filled-in fields could not easily be restored. And the income from container parks was far greater than the statutory fines.

This degradation also increased the instances of lowland flooding, through blocking innumerable channels. That rural people, once familiar with paddy streams, led the way in ruining the drainage system is surely noteworthy; that the very same people later blamed the government is telling. In fact, as the columnist Kevin Sinclair wrote in 1995:

> *But, remember! It is the original New Territories' residents who largely caused the smudge in their desperate urge for easy cash. They must share the blame, with the inept administration, for the deplorable state of the New Territories today.*

The New Territories' lowland streams, grossly degraded by pig and chicken manure, suffered from similar disregard. By 1987 Hong Kong's livestock waste was equivalent to the sewage from 1,600,000 people, and a ten-year programme was introduced that year to rid the polluted streams of manure. However, because of villagers' objections, the scheme was limited to certain areas, notably near Tolo Harbour. By 1993 the managed streams had shown a 60 per cent improvement in pollution levels—but the uncontrolled streams remained virtual cesspools.

Despite this sorry neglect of the old farmlands there were pockets of hope—such as Kadoorie Farm and Botanic Gardens. In 1951, when the Kadoorie Agricultural Aid Association was founded, the Farm's land near Tai Mo Shan's lower northern slopes was stark and treeless: today it is covered with shrubs and trees. Indeed, even inside the Farm's perimeter, some places are now so densely vegetated that they cannot be penetrated. The Farm also has a conservation area that supports countless species, while its ecological research brings new insights into Hong Kong's flora and fauna.

Education has always been the impetus behind Kadoorie Farm's vision. But today, in place of the farmers who found hope there in the 1950s, the Farm educates city people keen to enjoy, and learn about, Hong Kong's ecology and agricultural history: its flora and fauna, its farm plants and animals. Kadoorie Farm remains guided by ideals similar to those expressed over 2,000 years ago by the poet Chao Zuo:

> *Without good cultivation man has no bond*
> *That connects him with his soil.*
> *Without this bond he abandons too readily his place of birth.* *

* *The poem was quoted by Professor R. D. Hills in his Inaugural Lecture, 'To the Hills I Shall Lift My Eyes', given at the University of Hong Kong on 17 November 1994, which referred to land degradation and conservation.*

Coastal rubbishing is now an inescapable fact of the Hong Kong countryside. Here, near Sai Kung, there was more rubbish behind the camera than in front of it.

The villagers who not long ago farmed Hong Kong were imbued with such values—as the 300-odd *fung shui* woods that still grow around the Territory show. Few now have economic value, but their native trees still produce seeds—and saplings. These woods, invaluable repositories of botanical species, have enriched nearby forestry plantations—and so hastened the spread of mature woodlands.

In the higher country, almost all within Country Parks, reforestation has continued. Each year, government foresters plant hundreds of thousands of trees, mostly during the spring and summer. Many areas, reforested decades ago with single-species plantations, have now developed into mature, mixed-species woodlands.

By 1993 almost 10 per cent of the Territory was covered with woodland, or mixed woodland-plantation; about 5 per cent had plantations; and approaching 10 per cent had tall shrubland. Indeed, while other tropical countries have seen their forests much reduced since 1945, Hong Kong has seen its own much increased.

Three facts, however, detract from the Territory's impressive reforestation record. First, local industry is a major consumer of imported tropical hardwoods, thus depleting forests elsewhere. Second, some ecologists consider that there has been too much emphasis on the number of trees planted—and too little on choosing species likely to mature into ecologically diverse habitats. Third, fires caused by human disregard continue to exact a depressing toll—and so retard the development of more extensive woodlands.

The affluent Hong Kong of the 1990s highlights five key forces behind modern environmental degradation: rapid population growth; urbanization and industrial production; consumer consumption; escalating volumes of waste; and public complacency. In Hong Kong the processes have been concentrated in time and space.

Pollution haze, evident from any Hong Kong peak-top, occasionally takes on awesome densities. Here, seen from Ma On Shan, the sun disappears into a blanket of polluted air above Kwai Chung.

Since 1980 Hong Kong's population has increased by about 15 per cent, but its gross domestic product (GDP) has grown by a staggering 300 per cent. Such economic growth, allied with high population densities, has produced dire concentrations of waste. Hong Kong's 1994 total solid waste amounted to 8,300 tonnes per day, predicted to increase by 50 per cent by the year 2006. Landfills encroach on the countryside, but very little has been done to reduce the waste volume.*

In the countryside, only the accessible lowlands have been actually 'developed'. Pollution, however, knows no boundaries—and one cannot now ignore the impact of urban pollution on Hong Kong's upland country and remote coasts.

The local air, while better than in most Asian cities, is severely polluted. Emissions from industrial plants have been reduced gradually, but action to curb air pollution from vehicles remains mired in weak enforcement, few prosecutions, and projections of future growth. Statistics can numb. But no one could ignore the message of one afternoon I spent on Ma On Shan: forty minutes before sunset, the sun dropped out of sight behind a flat-topped, grey-black band—Hong Kong's atmosphere, the air we breathe daily .

Since the late 1980s there have been modest improvements in water quality, especially around the outlying islands and in Tolo Harbour. However, during the early 1990s Hong Kong was still fighting a rearguard action against decades of gross water pollution. The EPD's own 1993 review admitted: 'Black stinking mud characterizes the places most affected, and it doesn't take an expert to know something is wrong'. Meanwhile, with massive reclamations demanding very widespread dredging, the silting of marine habitats is causing extreme concern. Between 1990 and 1993 licenced marine dumping of landfill increased by *twenty* times.

Tonnes of rubbish drift around the coasts. On remote, otherwise paradisical beaches—from the far north-east to the remote south-west—flotsam fouls beaches and rocky shores alike. Much of this rubbish comes from boats and, given its composition, much of it from the fishing fleet. But barbecue prongs jabbed into the sand, cans and bottles dropped beside pathways, and much more—these are the rubbish of land dwellers.

Along remote beaches, often the most beautiful, it is not uncommon to see actual *mounds* of debris. Casual surveys indicate that synthetic materials account for most of the rubbish. On one distant Lantau beach I noted *plastic*—bowls, plates, mugs, ice cream basins, detergent bottles, soap holders, brooms, thongs, shoes, hats, coat hangers, shampoo bottles, shopping bags, soft drink bottles, shower curtains, petrol containers, nylon cord and rope, rubber inner tubes and tyres.... One 1994 newspaper report reiterated:

> *Plastic bags are an environmental headache, leaking toxins into the ground, choking marine life, spoiling Country Parks and beaches, and filling up dwindling landfill space. Plastic takes hundreds of years to disintegrate.*

Governor Christopher Patten (1992–) sought to give the environment a high priority at the start of his term. But past neglect compromised his aspirations; and the environmental task force, announced by the Governor in October 1993, and which aimed to clean up notorious 'black-spots', was faced with problems which had demanded urgent action at least a decade before.

Hong Kong's small area, much of it too rugged to develop, and its high population, have led to some of the world's highest population densities. In 1994 only about 16 per cent, or 170 square kilometres, of the Territory was 'urban'. On to this relative scrap of land were crowded some 6,150,000 people. As a result, the city has eaten ever further into its countryside.

The new airport at Chek Lap Kok illustrates this undeniable fact, for the project is the most massive single 'urban' intrusion to date into the local countryside. Impressive as Chek Lap Kok and its infrastructure no doubt will be, the airport's development will have had profound environmental impacts: on the natural landscape of north-west Lantau; on specific land habitats; and on the surrounding marine habitats.

To gaze down from Lantau's high peaks over the airport site underlines the first impact. Not even the most ardent advocate of the project could deny that the rural charm of north-western coastal Lantau is now lost. The island's uplands remain intact, for now, as do its eastern and southern coastal slopes. However, if the airport infrastructure is connected by road to Mui Wo and Tai O, Lantau will be doomed to spreading development—as large areas near the coast lie outside the island's two Country Parks.

A tiny frog illustrates the second impact. Romer's Tree Frog—green-brown, about the size of an adult finger nail—is believed to exist nowhere except on a few Hong Kong islands. Perhaps a quarter of the species' *world* population once lived on Chek Lap Kok: today, amid the awesome building site there, both habitat and frogs are gone forever.

The fate of a little-known dolphin emphasizes the third impact. The Chinese White Dolphin—in fact more pink or grey—lives in

* *The statistics on pollution given here almost all come from government publications, in particular the Environmental Protection Department's annual reviews.*

the waters north of Lantau. About a century ago there were perhaps as many as 400 of them swimming in Hong Kong waters, but by 1994 their population was estimated to have fallen to about 100. Dredging and underwater blasting have severely affected the water nearby, and destroyed many kilometres of coastline. Nothing can alter the fact that the airport, at least during its construction, has degraded the seas around Lantau. There are now real fears that the Chinese White Dolphin may become extinct locally.

As if this were not enough, Lantau is to be subjected to yet another massive development—the new Lantau Port. Barely debated in public, indeed with no real consultation with the Lantau people immediately affected, the proposed container port was confirmed by the government in mid-1995.

The development, centred at Penny's Bay along the north-east of Lantau, will have major environmental impacts. The beauty of the area will be compromised and the waters and tidal flow will be profoundly affected, since the width of the channel between Lantau and Hong Kong Island will be almost halved.

Both the airport and port are part of the Territorial Development Strategy (TDS), the government's umbrella plan for all future major developments here. Few ordinary people comprehend the scope of the TDS, few have any inkling of its gargantuan 'developments'. Indeed, as *One Earth*, journal of Friends of the Earth, observed in 1995 of the TDS:

> *These developments are intended to bring economic benefits that have never properly been defined, but they will also require trade-offs that have never properly been discussed.... [Yet] there is no public debate on whether this path is one worth travelling down.*

The TDS highlights a fundamental issue: how much economic growth and trade can Hong Kong absorb without forever ruining its environmental quality and natural landscape. Elsewhere the need for trade development to be 'sustainable' is being urgently addressed. But as Lisa Hopkinson and Kathy Griffin wrote in *One Earth*: 'Economic growth is Hong Kong's sacred cow, and strategies allegedly fuelling growth are rarely questioned.'

The ideological gulf that once separated Hong Kong from China gave the impression that topography—and even life—ended at the Shenzhen River. In reality, of course, Hong Kong has always been an integral part of South China, particularly of Guangdong Province.

The traditional Chinese ideal of living in harmony with nature was disrupted by the poverty of the decaying Qing Dynasty, the chaos of the twentieth century, and the Communists' post-1949 drive to 'modernize'. The resulting dramatic changes still sweeping across China have had—and will have—profound effects on its natural landscapes.

During the late 1970s Guangdong was one of China's poorest provinces: today it is the richest. In 1994 Guangdong produced almost 10 per cent of China's GDP—Shenzhen and Zhuhai (near Macau), on either side of the Pearl River estuary, were the country's two wealthiest cities—and migrant workers flocked to Guangdong. But, as elsewhere in China, relentless industrialization has come at a cost: large tracts of farmland and half-wild habitats have disappeared, and pollution has escalated alarmingly.

Hong Kong capital and enterprises have played a pivotal role in Guangdong's transformation. In the decade from 1981 to 1991 the value of Hong Kong's imports from China grew about nine-fold. In the two decades after 1991 estimates suggest that the total tonnage of Guangdong trade passing through Hong Kong will *triple*—from 100 million to 300 million tonnes. This trade, funnelled through tiny Hong Kong, will place phenomenal pressure on its land, and on its rural and urban environments.

Guangdong's economic 'miracle' poses three other distinct threats to Hong Kong's countryside. Development in Guangdong has affected the habitats of species that visit Hong Kong, especially birds; pollution has spread into Hong Kong through the air and rivers; and the development now consuming Guangdong may well spread into Hong Kong and consume its countryside also.

The modern destruction of wildlife habitats in South China began decades ago. During and after the Great Leap Forward of 1958 very extensive areas of forest were cut down for fuel, but only sometimes replanted. Recent development has had an equally severe impact, by turning farmland that previously held animal and bird populations into steel and concrete jungles. Mao Zedong, by re-interpreting the Chinese legend, *The Foolish Man Who Moved the Mountains*, inspired the masses to believe that they could 'move mountains'—and without help from the gods (as in the real legend). In so doing Mao bred a dangerous human arrogance towards nature.

The danger of pollution spreading in from China was brought home by some recent disasters. Most dramatic was the August 1993 explosion of toxic chemicals in Shenzhen. More insidious were the June 1994 floods. Besides causing great devastation and loss of life in Guangdong, the floodwaters flushed vast areas of

Massive land reclamations are now impacting severely on rural and urban environments. This Shau Kei Wan scheme covers the remnants of the last natural bay along Hong Kong Island's northern shore.

countryside—and swept worrying amounts of chemicals and organic matter into Hong Kong waters. As the *South China Morning Post* observed:

> *Environmental problems know no borders and the economic integration of Hong Kong and China in the past decade has driven home the message that China-originated pollution can affect Hong Kong and vice versa.*

However, perhaps the greatest threat from Guangdong is the danger of ill-conceived projects actually spreading into the Territory after 1997. Today the Shenzhen 'Special Economic Zone' presents a wall of concrete facing the Shenzhen River and Hong Kong's lowlands. Given the *ad hoc*, fragmented, and sometimes chaotic planning throughout the Pearl River delta, the northern New Territories could very easily become enmeshed in Shenzhen's expanding transport links and other projects. Cross-border government-to-government committees are currently addressing this issue, but what is agreed remains to be seen.

That many Mainland interests see Hong Kong as a treasure trove to plunder is true. And the danger of spreading 'infrastructure' was shown in December 1994, with the unilateral announcement by Zhuhai officials that they intended to build a bridge to Tuen Mun in Hong Kong—spanning 30 kilometres of the Pearl River estuary. The bridge, if it is completed, will have a major impact on the estuary, as well as on the Tuen Mun area. With initial work at Zhuhai already started, Hong Kong was treated to the brazen disdain of one Zhuhai official. As reported in the *South China Morning Post*, this official stated:

> *We are going to build a bridge from Zhuhai to Hong Kong. This is what we have decided. It is the will and deeply cherished dream of the 1.2 billion Chinese people to build a bridge from Zhuhai to Hong Kong. We will let you know where it will land when we finish.*

A glance at the map of Guangdong emphasizes three things. That Hong Kong's indented coastline is the most sinuous and beautiful section of the province's coast. That only too easily Hong Kong could be ruined forever by hasty, environmentally dubious schemes. And that Hong Kong is too small to absorb, without environmental losses, the trade of so large a powerhouse as the hinterland of Guangdong.

The countryside is increasingly popular in Hong Kong, as weekend hikers with Reeboks and Walkmans show. Yet the government seems out of step with this desire for 'natural' recreation, and with the growing 'green' awareness that it represents.

Since their beginnings almost twenty years ago the Country Parks have been greatly improved, and today offer still-wild areas and also good facilities. But many conservationists argue that the Parks need a new vision—and new directions. Above all, until the government boldly decides to take over the many disputed pockets of land near the Country Park borders, such as at Sha Lo Tung and Tai Long Wan, none will be safe from developers.

But that would demand generous financing, not the miserly amounts the Country Parks now receive. Together, the twenty-one Country Parks cover some 40 per cent of Hong Kong's land area. Yet the Country Parks' budget estimate for 1994 to 1995 was just HK$202 million—which provided a staff of about 1,350. This funding represents 0.147 per cent of the government's total appropriations for 1994 to 1995.

Given this woeful level of recurrent funding, it is hardly surprising that the government has refused to provide the additional modest funds needed to extend—or establish—some Country Parks. Annually, since the early 1990s, the Country Parks Board has requested action on the extension to the Lantau North Country Park, agreed in principle by government as a key mitigation measure to compensate for the loss of country to the new airport—and strongly supported by the airport consultants. There are also plans for a Lamma Island Country Park.

But in both cases the government's reply has been that it has insufficient reserves to find some HK$20 million for the establishment costs of these Country Parks. Yet the total cost of the Chek Lap Kok airport (excluding all its associated developments) is about 50 billion dollars—or about HK$27 million per day over five years. The figures speak for themselves.

With more than enough for Hong Kong nature-lovers to regret, what is there to preserve in the local countryside? Judging from some government planning, and the Environmental Impact Assessments (EIAs) that since 1987 have evaluated major developments, one might well conclude: very little. For until recently EIAs have been concerned much more with the human environment than with ecology and habitats; and they have mitigated, but not stopped, some questionable projects. It is surely revealing that the EIA for the Chek Lap Kok airport project was made *after* the site had been finalized and that the EIA ecological study for a reclamation that will engulf Green Island took just a single day.

Yet remarkably, despite its dense urban areas, the Territory still supports an extremely diverse flora and fauna, with more wild plant and animal species than in many much larger places.

The World Wide Fund for Nature's (WWF) *Ecological Database*, published in 1993, shows that Hong Kong has (approximately) 210 seaweeds; 175 ferns; 1,900 flowering plants, including 120 orchids; 2,000 moths; 200 butterflies; 93 dragonflies; 96 freshwater fish; 23 amphibians; 78 reptiles; 422 birds; and 57 mammals. Indeed, species unknown to science are regularly discovered, and much remains to be learnt concerning the vast majority of known species. As David Dudgeon and Richard Corlett wrote in their recent ecological study:

> *What we see around us, on the hills and in the streams, are the scattered pieces of Hong Kong's original biota.... [Yet] the terrestrial habitats in Hong Kong are, on the whole, better preserved than those in adjacent areas of mainland China. Conservation in Hong Kong is still possible and still worthwhile!*

Littering, despite the efforts of the Country Parks, remains an intractable problem.

Some species populations are very low, but others are high; and some, especially those of birds and small mammals inhabiting the spreading woodlands, appear to be rising. Significantly, field research by Dudgeon and Corlett, made since about 1990, revealed new woodland areas and species unknown (or only rarely seen) when Stella Thrower was investigating the countryside in the late 1970s.

However, this recolonization process is limited, partly because, with the destruction of habitats in China, many species have virtually disappeared from the region. Hence Dudgeon and Corlett suggest that, if Hong Kong's remaining countryside is to be ecologically 'restored', the deliberate introduction of some previously native species perhaps should be considered. They conclude:

> *Whether we like it or not, we are the stewards of Hong Kong's physical and biological environment—nature alone cannot heal the wounds that man has inflicted over centuries.*

A valuable step towards an ecological overview of the Territory was the publication of WWF's *Hong Kong Ecological Database* (quoted above), and its accompanying extremely precise vegetation maps. Developed with support from the Caltex Green Fund, the work marks a break from past studies of Hong Kong land use, which generally stressed human activities over ecological habitats.*

The *Ecological Database* points to an issue too often ignored in the past: that safeguarding Hong Kong's species diversity demands more than merely drawing boundaries around obviously important areas such as the Mai Po marshes or Sha Lo Tung valley, or around sites known for specific, rare flora and fauna. The entire natural landscape has habitats and species of value.

The *Database* also highlights an inescapable fact: given Hong Kong's size, its countryside cannot be seen in isolation from the urban areas. Indeed, except on the outlying islands and in the most remote parts of the New Territories, Hong Kong's uplands are now hemmed in on almost every side by urban and semi-urban areas—and by creeping development.

Excluding cross-border issues, and the power and speed of computer-planned–juggernaut-driven development, probably the greatest threat facing Hong Kong's countryside today is the local attitude to 'change' and 'growth'. For 'change' and 'growth' have become so pervasive that they are now the accepted norms: another tower block rising here, another hillside destroyed there.

* *The wide range of people who contributed to the Ecological Database indicates the expertise and commitment available to support the countryside.*

With this ingrained mind-set, planners too often extrapolate from previous 'growth' statistics, to envision awesome projects—or, blind to the cumulative results of incremental change, they sketch in yet another 'insignificant addition'. It sometimes seems that the sheer challenge of overcoming—of mastering—the local topography has become an end in itself.

The urban ecologist C. Y. Jim recently observed: 'Hong Kong is one of the few big cities in the world with, close at hand, an excellent natural scenic backdrop'. But that, he added, 'is unfortunately not much realized, even locally'. The peaks which frame Victoria Harbour—always a tourist mecca—are integral to its visual drama. Yet, blind to that fact, developers are erecting such soaring towers that they are actually blocking out even distant views of Hong Kong Island and the Kowloon hills. And, inconceivable as it seems to many, government development plans include the *destruction* of Green Island, just off Kennedy Town. This lovely island, the last of the harbour islands, is scheduled soon to become a hillock surrounded by reclamation—and, like Stonecutters Island already, engulfed by 'development'.

Countryside conservation is further compromised by the often fragmented nature of government decision-making, and the seeming inability of Hong Kong officials to make bold, 'anti-development' policy. As David Melville, Executive Director of WWF, wrote in 1993: 'Government still lacks a comprehensive conservation policy which brings together Country Parks management, town planning and wildlife conservation as a whole.'

Numerous proposed 'developments' have shown starkly the threat of projects slipping through one or another department or legal loophole and so impacting on the countryside.

In 1994, for example, there were commercial plans to develop some areas immediately adjacent to priceless countryside. Yet these places, often originally excluded from Country Parks because of the significance to villagers of their lowland streams, fields, and *fung shui* woods, have particularly valuable habitats and ecosystems. At present such localities are, as Dudgeon and Corlett wrote, 'protected by isolation rather than by law and are thus always vulnerable to future development'.

Hong Kong is a signatory, through Britain, to the key international agreements which aim to enhance and sustain the world environment.* However, although the government is committed

* *These international agreements include the Ramsar Convention, the Convention on International Trade in Endangered Species of Wild Fauna and Flora (CITES), the Bonn Convention, and various other conventions governing environmental quality. The Territory is also a signatory to the principles drawn up in 1992 at the Rio de Janiero Earth Summit.*

to implementing these conventions, the record of the 1980s and early 1990s indicates that its aspirations often exceeded the reach of legislation, administrative power, and funding.

The illegal dumping on old paddy field wetlands illustrates the gulf between government 'commitment' and action. The practice, though obvious to anyone and widely considered as detrimental at a policy level, went unchecked for over a decade. Meanwhile, the departments concerned had insufficient funds (and probably insufficient will) to address the legal loopholes—or to fund rapid, effective on-the-ground enforcement to prevent the wetlands' destruction.

Yet the government cannot be blamed entirely for the absence of more vigorous action. Indeed, as Cecilia Chan and Peter Hills conclude in their study of Hong Kong's grassroots environmental attitudes:

> *Until the latter part of the 1980s, government initiatives in the environmental protection field were typically launched against a backcloth of public apathy, and opposition, sometimes intense, from industrial interests.*

Education is now of critical importance in preserving Hong Kong's countryside and wider environment. Local people recoil from the degradation they see in Guangdong, and most admit to at least disquiet concerning Hong Kong's own urban pollution. Yet as Governor Pattern lamented at the opening of the Legislative Council's 1993 session: 'For all our achievements [we have failed] to mobilize the support of our community behind our efforts to clean up the environment'.

The Territory's leading 'green' groups—WWF, Friends of the Earth, Green Power, and the Conservancy Association—are in the vanguard of the drive to create a new environmental ethic. But, marked as their successes have been, only about 15,000 people—or one person in 400—belong to one or other of the groups. This is a fraction of the figure for other 'developed' countries.*

However, by highlighting undeniable environmental woes, and through initiating environmental and ecological research, the Hong Kong groups have become widely respected. Today they represent far more than mere fronts for 'green' lifestyles—they have far greater influence than the size of their membership might suggest, and are well represented on bodies such as the Advisory Council on the Environment (ACE).

* *During the late 1980s concerns about the Chinese nuclear power station then being built at Daya Bay, 50 kilometres from the centre of Hong Kong, hastened the development of a 'green' consciousness—as did the direct elections after 1985 to Hong Kong's various local government bodies.*

One of the groups' achievements has been their role in influencing the private sector, whose economic and political power is enormous. Many companies now have active policies to enhance their own environmental 'housekeeping'. However, as yet there is little evidence of serious debate in the business community over the impact of continued 'growth' on the natural landscape or of Hong Kong businesses addressing the wider issues of 'sustainable development'. Indeed, as with ACE, the government environmental advisory body, there is a reluctance to challenge the *status quo*—to rock the Hong Kong boat.

Executive, a magazine reflecting business interests, in an outspoken editorial in 1995 stated that a 'more activist approach' was needed—and noted that many companies still regard the wider environment as an issue for 'selective charity'. The editorial concluded: 'Corporate Hong Kong should look to its conscience.'

There are some notable exceptions—companies which have on their own initiative already resolved to make a contribution to cleaning up Hong Kong and to preserving its countryside.

The Swire Group led the way when it financed the building—and recent extension—of the Swire Institute of Marine Science at Cape D'Aguilar. The Institute, opened in November 1990, is staffed by University of Hong Kong marine scientists and coordinates research into marine ecology and degradation. When Swires developed the Pacific Place complex, the Group invested very heavily in preserving a venerable Chinese Banyan which had been planted on the site 120 years ago. Swire's decision was, as C. Y. Jim wrote, 'a pioneering example in Hong Kong of a building adjusted to accommodate an existing tree'. Swires are also leaders in the Private Sector Committee on the Environment and fund various scholarships. The Caltex Green Fund and the Woo Wheelock Green Fund are two generous environmental funds, which encourage a range of valuable initiatives.

Private sector funds also have helped establish three countryside field studies centres—at the Mai Po Marshes Nature Reserve, Kadoorie Farm, and the Lions Nature Education Centre near Sai Kung. Nature, experienced and learned through actual encounters, is their common theme. Given the Hong Kong educational emphasis on curricula and exams, such experiences are critical for fostering conservation values.

Environmental studies are also being expanded at every level of education, though the process is compromised by the traditional Hong Kong emphasis on 'textbook knowledge' over 'discussing attitudes'.

The first Environmental Resource Centre for community groups and schools was opened in December 1993. The centre, sited in the old Wan Chai Post Office and funded by the EPD, is a very modest beginning. Indeed, in a city as wealthy as Hong Kong,

Much of Hong Kong's reclamation fill comes from seabed dredging, which creates widespread silting of the surrounding waters—as here, near Shek O. Silting impacts severely on marine life.

one must wonder why this is the *first*—not the *tenth* or *twentieth*—environmental resource centre. None the less, it is a start, a small beacon of hope amid Wan Chai's polluted canyons. This Chinese couplet frames the entrance:

> *If we foul our world that sustains us, what then shall we eat?*
> *If we scorn hygiene that protects life, where then shall we live?*

Gleaned from the press between 1993 and 1995, while I completed this book, were signs of hope for the Hong Kong countryside—and signs of despair. There was evidence that Hong Kong had finally become a rooted community that cared for its soul—and so for its surroundings. Yet there was also evidence that Hong Kong remained a transient place where the pursuit of money was all consuming—where the vast majority gave only lip service to preserving the environment.

Newspaper reports also revealed the need for open communication and public participation, if the delicate balancing of interests necessary to preserve the countryside and improve the urban environment—and yet still sustain prosperity—is to be achieved.

In October 1993 the government admitted that the north-western New Territories' paddy fields, by then mostly illegal dumps, could never be restored to 'green-belt' appearance. Tony Eason, Secretary for Planning, Environment and Lands, stated that the promised 'clean up' of the area was likely to result in 'controlled urbanization' not 'green fields'. He cautioned:

> *If we do not recognize the market demand for these facilities, the people will simply go elsewhere, and we will find ourselves chasing them all over the country.*

Conservationists, and some members of ACE, regularly expressed concern over the implementation of EIAs. EIAs, they argued, were often seen as mere palliatives to 'soften', but not halt, projects. Many claimed that EIAs were hasty or inadequate, some that they were veiled in secrecy. Without doubt, EIAs were often seriously flawed—and sometimes mere window-dressing.

However, David Melville, of WWF, stated in his 1994 report that

Kowloon, once a narrow peninsula, now spreads far into Victoria Harbour. Whether all the present and planned harbour reclamations are justified is at least debatable. Much less uncertain is that if the Territory further degrades its urban environment it is less likely to conserve its surrounding countryside.

there was increased consultation with the 'green' groups concerning EIAs—especially for Deep Bay and Mai Po:

> *It is heartening to see an increasing awareness of the need to conduct ecological studies for these reports which, traditionally, have principally addressed issues relating to air, noise and water pollution at the expense of flora and fauna.*

During 1994 the Post Office issued four stamps of local corals: brilliantly coloured, they were shown in translucent aquamarine water. The reality was different: even around the outlying islands the water was murky from pollution and dredging. Near the inner harbour the situation was far worse. Early in 1994 the EPD admitted that making Victoria Harbour a Water Control Zone was 'a pointless exercise': with some two million tonnes of untreated effluent still entering its waters daily that was surely true.

In 1994 the government at last announced that a Marine Reserve and two Marine Parks were to be gazetted, the former at Cape D'Aguilar, the latter at Hoi Ha Wan and Double Haven. Partly because of funding limits these protected zones, first mooted in the late 1980s, covered only very small areas of inshore water. Marine scientists welcomed the 1994 announcement, but from the fishing fleet there came renewed, strident opposition. These marine conservation areas were finally approved in May 1995.*

Concern about the Chinese White Dolphins in Hong Kong waters was translated in 1994 into a three-year research study based at the Swire Institute of Marine Science. But the study came late indeed: for by 1994 the worst impact of the Chek Lap Kok airport's site formation had already occurred. In March 1995 ACE reversed its previous criticism of the Sha Chau fuel depot EIA—thus further compromising the dolphins' habitat.

International concern over pollution in China led to pledges of 'clean up' funds from the World Bank and elsewhere. But it was uncertain whether or not the Beijing authorities could enforce environmental responsibility in the provinces, notably Guangdong. Concerning Hong Kong, Chinese officials and environmental bodies seriously questioned the wisdom of the harbour reclamations and the Lantau port plans.

The fact that the old paddy fields were still commonly seen as 'worthless swamps' led to a major ecological study, which began in January 1994. The ecologists intended to identify Hong Kong's most valuable remaining wetlands. One of them, David Dudgeon, commented to the media: 'The environment we have out there is the only one we've got. When it's gone we can't make another one.'

* *Following the approval of these marine conservation reserves, the Country Parks Board was renamed the Country and Marine Parks Board—with similar changes throughout the Country Parks system.*

Hong Kong shows a most marked degree of community generosity in many spheres—and countless individuals work willingly for the common good. Yet in late 1994 a newspaper survey reported that 60 per cent of respondents considered money 'more important' than anything else. The newspaper warned:

> *We are in danger of remaining a first-class economic entity with a third-class mentality, better suited to places still fighting for mere survival and incapable of striving for higher goals.*

Mei Ng, Director of Friends of the Earth, commented in 1995: 'Hong Kong people have that borrowed place and borrowed time mentality. They are still only living for today.'

Understandably, as 1997 approaches Hong Kong's people are increasingly preoccupied with the transition. Concerns about the future are leading a minority to new homes elsewhere—and the majority to attempting to maximise short-term gains here. Thus the Territory displays both its strength, vigorous individual enterprise—and its weakness, transience and a resulting lack of responsibility.

Hong Kong has been almost totally transformed from its primeval origins. Yet, clever as mankind may seem, we are still unable to control much that defines existence. Foretelling what nature will do is fraught with uncertainty.

Who could have said in 1993 that the following year would bring to Hong Kong and Guangdong the wettest July on record? Or that the mean temperature for the following December would be a clear two degrees Celsius above the usual?

And what of the 'seabed kill' that occurred in Mirs Bay, along Hong Kong's north-eastern shores, during the previous summer—when almost every coral and marine organism living below two metres deep suddenly died in July 1994.

The exceptional rainfall that fell that month certainly contributed, as did the resulting massive discharge of water from the Pearl River and local streams. Nutrient-rich pollutants may, or may not, have been involved. All that is certain is that, beneath warm well-oxygenated surface water, colder oxygen-depleted water had drifted in from the oceanic depths. Trapped below the surface water, the colder water starved the organisms of oxygen—and killed them.

But what led to—or triggered—the ocean inflow itself? There are various theories, but no definitive answers.

230.4
EV 3799

Dawn at Mai Po. Sunrise colours the sky behind the Lam Tsuen hills, casting a sheen over a Mai Po pond.

REGION

THE MAI PO WETLANDS

In the Hong Kong context, 'wise use' of wetlands all too often seems to equate to the 'filling in' of wetlands for development.... A filled-in wetland is no longer a wetland and thus, filling in wetlands cannot be considered 'wise use'.

LEW YOUNG, 1994

It is midwinter. I am just ten metres above sea level. Anywhere else in Hong Kong such an elevation would be insignificant and unremarkable. But in the north-west of the Territory, at the Mai Po Marshes Nature Reserve, those few metres matter: they give a commanding view over the surrounding flat wetlands.

From a police border lookout tower, I gaze out into the pre-dawn gloom. The eastern sky is slowly turning to fire, enveloping the marshes spread out below in a ruddy glow. Cormorants and egrets rise with the brightening light, flapping and wheeling as they gain height.

Behind me, Deep Bay's mangroves and muddy waters are slowly taking form. On either side, to the north and south, golden light touches the massed buildings a few kilometres away: across the border in Shenzhen, and in Hong Kong at Yuen Long and Tin Shui Wai. Nature seems hemmed in by development.

Yet only twenty years ago the area was mostly agricultural, and when G. A. C. Herklots explored for birds here before the war it was entirely rural. The land bordering Deep Bay was then ideal both for bird-watching and for shooting wildfowl, as its habitats were so varied: *fung shui* woods, banana and fruit orchards, sugar cane, maize fields, rice paddy, shrimp ponds, mangrove swamps, and mud-flats. Indeed, it was a place where people seemed to live in harmony with nature.

The most unusual habitats for Hong Kong, and those with the most birds, were the mud-flats and mangrove swamps lining Deep Bay. There, paddy water rich with pickings for birds drained out along man-made dykes to meet Deep Bay's salt-and-silt water rich with fish fry, tiny shellfish, and other marine organisms. Hardly surprisingly birds, especially waders, congregated there in great numbers.

Herklots was no doubt a familiar figure to the local villagers. Time and again he would have been seen striding out along the bunds, binoculars and notebook in hand:

The expanse of fields intersected with dykes and swamps always yields a number of waders and water-loving birds, and I saw eight species before reaching the bund. They were a White-faced Wagtail, four or five Eastern Great Egrets, a Marsh Harrier, the first of three, a Cormorant on the water, a few Cattle Egrets, the first of many Little Egrets, the first Sandpiper, and six large Terns, probably Caspian Terns....

[Beyond the bund were] two Purple Herons, rare birds which are difficult to see in marsh or mangrove swamp, a Fantail Warbler in the mangrove, a Redshank on the bund, a Snipe and then three unidentified birds.... A Kingfisher and four Little Grebe completed the birds seen from the bund, and brought the list up to forty-five.

It was Easter Monday, 19 April 1941. By late that afternoon Herklots had recorded sixty-nine species—the highest pre-war bird count made in a single day in Hong Kong.

Deep Bay and its coastal wetlands are like no other place in Hong Kong. Nowhere else is the water so turbid and silt-laden—and nowhere else is the land so flat or clayey. Many other parts of the Hong Kong countryside are more majestic, but perhaps nowhere else is as ecologically valuable as this north-western corner.

Deep Bay's coast has Hong Kong's only extensive mangrove swamps, and is rich in flora and fauna. Most especially, numerous resident, visiting, and migratory birds make the Mai Po marshes a wetland of world significance—and one of the most important wildlife sanctuaries along the South China coast.

Yet this is far from a 'natural' landscape. Indeed, the wetlands around Mai Po are the result of centuries of toil by villagers who reclaimed them from extensive mangrove swamps. The agricultural landscape they thus created also suited nature, and especially waterbirds. That cannot be said for recent human impacts on the Deep Bay environment: for, in little more than a decade, developers have surrounded the Mai Po marshes with highways and urban clusters—especially at the New Towns of Yuen Long and Tin Shui Wan, and around Shenzhen. Pollution has long since spread through and around the bay.

The Mai Po Marshes Nature Reserve has more bird species than any other place in Hong Kong, and some that are threatened with world extinction. Yet, despite that, the Nature Reserve is beseiged, and its future is far from secure.

River silt forms the visible surface of the Mai Po landscape. Underlying the area is marble bedrock, laid down some 300 million years ago; and above that lies colluvium (eroded rubble) washed off the surrounding ranges. Much later alluvium, brought down by the post-Ice Age rivers, flooded over the Yuen Long basin and transformed it into today's alluvial, clayey plain.

The silt that gradually formed these north-western lowlands came partly from the Shenzhen River and smaller local streams. However, the far mightier Pearl River was the main source, as tides washed its heavily silt-laden water into Deep Bay.*

Mangroves later established themselves across this alluvial plain, as no other plants could survive the salty environment caused by high tide inundations. The trees, whose roots naturally trap and hold silt, aided the process of sedimentation.

The exact ecological picture before the Deep Bay area was first settled is uncertain. However, it is likely there were six species of mangroves (as today), and there was certainly a wide diversity of marine species that favour brackish water. Water snakes, and Long-nosed Crocodiles, and numerous waterbirds no doubt fed off the low-tide mud-flats—which were rich in small organisms such as worms, shrimps, shellfish, crabs, and mudskippers.

Settlement of the Deep Bay–Yuen Long alluvial plain paralleled the general waves of Chinese migration to Hong Kong. After the Song Dynasty settlement, farming took the place of fishing as the predominant activity around Deep Bay. Towns had developed at Yuen Long and Kam Tin by about 1300, when agricultural land was increasingly scarce.

* *The Pearl River and its tributaries drain some 425,000 square kilometres of South China. The river's annual flow today is reckoned to be about 308 billion cubic metres—or, averaging seasonal flows, abut 9,000 cubic metres per second. The river brings down almost 80 million tonnes of suspended sediment each year—or about 2.5 tonnes per second..*

So began the weary labour of draining and reclaiming the coastal mangrove swamps for farm land, while the mangroves were also cut back for coastal access and fuel. The process of reclaiming land was probably little different from that recorded by J. H. Stewart Lockhart at Deep Bay in 1898:

> *Walls of rough rubble backed by earth are constructed to keep out the sea water. Along the top of these embankments there is a footpath; and sluices, made of timber planking sliding in groves cut into the stone side walls, enable the villagers to keep out the sea water, and get rid of their surplus fresh water at low tide.*

Once drained of sea water, the swamps were burnt, cleared, and then flushed with fresh water to reduce their salt content—which often took a number of years. Only then could the land be planted with *haam moon*, 'brackish water' rice. Other local activities included evaporating sea water for salt, making bricks from local clays, and fishing.

The remaining mangroves helped both fishermen and farmers. The mangroves created a barrier against typhoon waves, so reducing the danger of flooding. They provided firewood, charcoal, medicines, tannin, and poisons for stunning fish. And, above all, the mangrove swamps were invaluable feeding grounds for the larvae and fry that sustained Deep Bay's fish and shrimp stocks.

Over the centuries, the mangroves that once surrounded almost all of Deep Bay were extensively cut back. However, though this man-created fishing-farming environment was much changed from its primeval state, it still provided varied habitats for land, river, and sea fauna.

When the New Territories became part of Hong Kong in 1898, the Deep Bay landscape was very similar to that of past centuries. 'Bamboos and banana plants fringe the mud-walled villages that are rarely out of sight of one another so intensely is this district cultivated', one European traveller wrote. Large areas of mangrove swamp had long since been reclaimed, especially around Ping Shan.

Deep Bay's coastal ecology remained economically important, as Rudolf Krone had reported in 1858: 'Hundreds of old men, women, and children may be seen on the extensive flats left by the receding tide, collecting the small fishes, crabs and other animals which have been stranded. The able-bodied men are with their boats at sea.'

Over two generations later, in the 1930s, G. A. C. Herklots passed 'mud and rice-straw thatched huts inhabited by families who live a rather miserable existence tending ducks, cultivating

Soon after sunrise birds gather at one of the Mai Po wetlands' many water channels.

Six mangrove species are found at Mai Po, each one colonizing an ecological niche suited to its salt-water tolerance.

rice and fishing in the dykes'. Herklots added: 'Below the village is a pond into which all the refuse gravitates.... On the top of the bund are bundles of mangrove roots grubbed up from the swamp to serve as fuel.'

Not only the local inhabitants exploited the area's rich ecology. R. C. Hurley had arrived in Hong Kong in 1879, and by the early 1900s he was the epitome of a local outdoorsman. Shooting excursions to the Mai Po area were among Hurley's favourite pursuits:

> *The country for many miles around is flat and very marshy, inhabited by many varieties of wildfowl in great numbers which find their breeding and feeding grounds in the brackish waters of a shallow bay known as Deep Bay.... Being tidal this bay is drained of its waters twice every twenty-four hours, exposing a surface of rich alluvial mud in which are very extensive oyster beds.*

Hurley generally took his fowling parties to Deep Bay by an overnight boat. At daybreak—and with 'a substantial tiffin to be consumed *en route*'—they went by sampan to the head of a creek:

> *Shooting begins immediately with waterfowl: curlew, plover, duck, teal, and sometimes geese. Snipe are always to be found from August until March, and spring snipe even later. Having secured the quantum of such as the river, marshes and paddy fields have to offer, we must now strike inland, making for the nearest village, Mai Po—where we may, if the season is well advanced, add a few pigeons to the bag.*

What J. C. Kershaw thought of R. C. Hurley's activities can only be guessed at. Both men lived in Hong Kong in the early 1900s, but they had very different interests: while Hurley was off shooting birds, Kershaw was out studying butterflies.

Hong Kong is remarkably rich in butterflies. The Mai Po area alone has over fifty species today, the most common being Orange Tigers, Blue-spotted Crows, Common Grass Yellows, and Common Black Jezebels.

Reeds are Mai Po's second most common plants. Once used in various ways by farmers, the reeds are no longer harvested and some spread rapidly.

Massed cormorants: hundreds of these visiting birds were perched nearby during this spring mid-afternoon.

As I see them, Hong Kong's butterflies flit around the memory of Kershaw—and through the pages of his *Butterflies of Hong Kong*, as beautiful as the insects it records. The book, published in 1907 and brought to life with brilliant plates, was the first detailed record of Hong Kong's butterflies. The work reflected determination, knowledge, and passion.

For seven years Kershaw devoted his spare time to collecting butterflies—through summers 'rain-swept and sun-burned'—across country 'clad with coarse grasses'—among 'a poor and dense population'—watching, studying, netting butterflies—'gorgeous and striking genera that fly hither and thither at a speed which quite eclipses their English relatives'.

The butterfly numbers varied greatly from year-to-year. Kershaw believed this was partly due to 'the continual destruction of the [larvae's] foodplants by the native fuel-gatherers', and to the destruction of vegetation by goats and pigs. However, though village children collected crickets, cicadas, and grasshoppers—mostly to feed to caged birds—they left butterflies alone. Indeed:

> *When out with the net, if villagers do not consider one simply mad, they sometimes ask if the butterflies are to be used for medicine.... One constantly hears 'Tseuk-tsai, tseuk-tsai'—'Small birds. He's catching small birds.'*

Back at home, the butterflies had to be 'set'—pinned and conserved for later study. Hong Kong's climate complicated the work: 'Set insects in this climate need an enormous amount of care and attention, and even then deteriorate rapidly. Damp, mould and mites play havoc with them in the wet season; in the dry months the antennae and legs snap off at the slightest touch.' So Kershaw sent his specimens 'home' as soon as possible: 'The butterflies can then be relaxed and set, years after they were captured if necessary, in as good condition as when first taken.'

So much effort, so much care: wooden drawer after wooden drawer of lovingly set, superbly coloured butterflies. There was, for example, the lustrous indigo-to-violet Blue-spotted Crow—known for its 'rather indolent floating or sailing flight'.

Deep Bay is Hong Kong's largest estuarine area. Its average depth is just three metres, and nowhere is it deeper than six metres; yet it covers 112 square kilometres, greater than the area of Hong Kong Island. Deep Bay's mud-flats, mangroves, and fish ponds form an interrelated, complex ecosystem.

The dwarf mangroves that still line a few kilometres of coastline are the sixth largest mangrove community along the South China coast. Six species grow there, each one colonizing a section of the foreshore—and thus providing a specific habitat for the varied marine organisms. These include the larvae and fry which sustain the Deep Bay fish and shrimp populations—and which make the mud-flats an invaluable bird feeding ground.

Reeds at the Mai Po Nature Reserve.

One of the many water channels that criss-cross the Mai Po Nature Reserve.

The other most common plants around Deep Bay and Mai Po today are reeds and sedges, especially the reed *Phragmites communis*. Previously the reeds were held in check by farmers, who used them mostly for thatching—and also for bindings, rain cloaks, plaited ropes, woven mats, and fuel. But with these all now replaced by synthetics the reeds have no economic use, and so are rarely cut.

Deep Bay holds armies of worms, crabs, oysters, shrimps, mudskippers, grey mullet and other fish, crustaceans, and small organisms. One small crab is unknown beyond Mai Po. The richness of Deep Bay's marine ecology led to shrimp-breeding in *gei wais*—earth-walled ponds constructed on old mangrove areas. By periodically flooding or draining them with the tides, shrimp breeders stocked their *gei wais* with shrimp larvae from the bay—and enriched their ponds with organic nutrients. To harvest the mature shrimps, the ponds were drained through sluices, when their muddy beds provided rich pickings for waterbirds. A similar, though less ideal, symbiosis occurred when, after about 1960, many of the shrimp ponds were turned into fish ponds.

Today, the food supplies yielded by old *gei wais* and fish ponds are a central reason for Deep Bay's many thousands of wintering waterbirds. Add to this the fact that around Mai Po (and throughout the New Territories) the wetland habitats based around paddy farming have disappeared in a few decades—and one appreciates why the Mai Po marshes are so important for sustaining the populations of waterbirds.

Over 320 bird species have been recorded around Deep Bay and Mai Po, as Lew Young and David Melville record. Most of these are considered to be common, but twelve are thought to be 'rare' or 'threatened'. Of these, five species seen at Mai Po represent at least 5 per cent of their world populations: Oriental White Storks, Black-faced Spoonbills, Asiatic Dowitchers, Spotted Greenshanks, and Saunder's Gulls. It is surely sobering that, based on birds wintering at Mai Po between 1994 and 1995, the Black-faced Spoonbills seen there accounted for no less than a quarter of the species' world population.

The significance of Deep Bay's bird grounds extends far beyond Hong Kong. Over 220 of the species recorded there are migrants or

A lone wader symbolizes the threats facing Mai Po. Can its wetlands survive? Or will relentless development, such as that of Shenzhen in the background, destroy them?

The Mai Po Marshes are surrounded on almost every side by development—existing, imminent, and proposed. This southerly view looks towards Tin Shui Wai New Town.

winter visitors which come to the bay in search of food: in 1994 and 1995 some 20 to 30 thousand migrants passed through, and some 60,000 visitors over-wintered there. The rich estuarine and wetland ecosystems provide the abundant foods that migrant and visitor birds require, allowing them rapidly to develop sufficient fatty reserves for their long onward journeys. From Mai Po some birds fly about 2,500 kilometres to north-eastern China, or around 4,500 kilometres to north-western Australia, without feeding *en route*.

The spring migration through Mai Po presents a remarkable spectacle, with countless thousands of waterbirds feeding across the bay's exposed mud-flats—and moving in towards Mai Po itself with the advancing tides.

❦

It is a few hours after dawn, and I am photographing from one of the many boardwalks that criss-cross the Mai Po Marshes Nature Reserve. A breeze rustles the tufted reeds that stretch along some nearby *gei wai* channels. Beyond a vista of trees, hills a few kilometres away frame the tower blocks of Yuen Long New Town.

Surrounded by urban developments, set amid much older reclaimed wetlands, the Nature Reserve is undeniably a man-made landscape. Yet, fundamentally, it is the preserve of birds, as marks across the boardwalk hint at: clayey, three-pronged patterns, the unmistakeable tracks of wading birds.

The Mai Po Marshes Nature Reserve has its origins in the Nature Conservation Area gazetted by the government in 1975, and which in 1976 was declared a Site of Special Scientific Interest (one of fifty such Hong Kong sites). Listed under the Wild Animals Protection Ordinance, the Nature Reserve has been managed since 1984 by the World Wide Fund for Nature Hong Kong. The WWF, by gradually raising funds, purchased old *gei wais*, and in 1993 the government provided HK$16 million to finance the purchase of the remaining shrimp ponds. The Reserve now covers 380 hectares—about half of them mangrove swamps and the rest old *gei wais* or fish ponds.

The Nature Reserve has three primary aims: to maintain, and if possible increase, the Deep Bay area's species diversity; to

The road leading into Mai Po passes polluted streams and general waste, such as these discarded truck tyres.

encourage environmental awareness through educational visits and projects; and to further local research.

The ecological research conducted at Mai Po is undoubtedly significant, while the 35,000 students and visitors who each year explore the reserve attest to its vital educational value. The number of bird species and their populations are also increasing. Three reasons account for this: the total ban on capturing local birds in Hong Kong since 1981; the destruction of neighbouring wetland habitats in China, forcing birds to go elsewhere in large numbers; and the enrichment of the Deep Bay food supplies due to increased organic pollution. Herklots' pre-war April record of sixty-nine species sighted in a day compares with today's typical counts of upwards of 100 species seen on April days.

The Mai Po wetlands, sandwiched between the relentless development of the north-western New Territories and Shenzhen, are among the most threatened countryside in Hong Kong today—as walking to Mai Po indicates.

The Sheung Shui–Yuen Long highway runs through overgrown farm land, now the scene of light industry, car wreckers, and container dumps. The Mai Po turn-off is marked by the forlorn remains of a once-gracious village complex, now just another container yard. Beyond there the dirt road to Mai Po winds along a causeway between duck and fish ponds. Many of the streams, clogged with plastics, are foul with putrid ooze. Along the way are the signs of uncontrolled sprawl and waste: old tyres, rusting containers, derelict machinery, abandoned cars grown over with creepers.

In the Nature Reserve itself, provided one keeps a low gaze, all is sylvan: ponds, mangroves, reeds, birds. But if one looks higher, in almost every direction major buildings stand only a few kilometres away. To the north, Shenzhen's wall of buildings rises above the wetlands; to the east, construction cranes lean like storks over a new residential complex; to the south the massed housing blocks of Yuen Long and Tin Shui Wai rear up above the flat landscape.

In 1898 only about 23,000 people lived across the Yuen Long plain, the hinterland of Deep Bay: by 1961 the population had roughly quadrupled. Evolutionary agricultural development ended about two decades ago. Since then the Mai Po wetlands have been engulfed by extremely rapid urban development and ever-spreading pollution.

Housing estates and transport infrastructure have dominated the development. In 1975 Fairview Park was begun: a low-rise housing complex for 30,000 people, it was built on ground reclaimed from abandoned agricultural wetlands. Fairview Park's area today is half that of the Mai Po Nature Reserve, onto which it abuts. As Fairview Park was being built the government was proceeding with its ambitious New Town housing programme. Yuen Long was formally designated in 1978, and by 1990 its tower blocks held some 120,000 people—and had a projected population of about 140,000 by 1995. In 1987 the construction began of Tin Shui Wai which looks onto the mangrove marshes—and now houses some 100,000 people.

North of the Shenzhen River the development was still more striking. In a single decade the market town of Shenzhen became the city of Shenzhen, and industry, housing, and recreation engulfed its surrounding agricultural land. Major transport routes now skirt Deep Bay, and, on its western arm at Shekou, port facilities have been greatly expanded. Meanwhile, in 1995 work was proceeding to deepen and widen the Shenzhen River—thus perhaps degrading a significant area of mud-flats near the river's mouth.*

The transformation of Shenzhen is indicative of the wider development that is affecting Guangdong, and which has destroyed many other semi-natural habitats—especially along the Pearl River. The development is *ad hoc* and fragmented, so greatly increasing environmental degradation and habitat losses.

Aerial photographs taken in the late 1970s emphasize how fast has been the change surrounding these Hong Kong wetlands. Then, the view over Mai Po was across an expanse of fish ponds which stretched into China. The plain, patterned with shades of green, was dotted with small villages and *fung shui* woods. Around Mai Po there was not a single modern building.

The development around Deep Bay has fundamentally changed its physical landscape, and greatly reduced its wetland habitats. Given the concurrent massive increase in the population, only the strictest controls could have prevented environmental degradation and pollution. However, the waste management controls were half-hearted at best.

Agricultural wastes, pesticides, herbicides, domestic sewage, and industrial effluents account for the present critical state of Deep Bay. The agricultural waste is mostly from pig and chicken farms. Much of the waste reaches Deep Bay from the grossly polluted streams of the Yuen Long plain and the Shenzhen River—and the rest from the Pearl River.

The figures are grim. Even in 1987, two-thirds of the water samples taken from Inner Deep Bay had less than 50 per cent oxygen saturation, and half had less than 10 per cent saturation. The concentration of faecal bacteria in Deep Bay water was then also extremely high, and oysters harvested from the Bay were—and still are—often found to be contaminated with heavy metals. Nitrogen concentrations in the water were also very high. Indeed, the steadily worsening pollution of Deep Bay was a major reason for the decline of its oyster, shrimp, and mudskipper fisheries.

In 1990, China's National Environmental Protection Agency described the pollution of the Pearl River as 'comparatively serious'; the river was then taking massive amounts of industrial and domestic effluent. And the Shenzhen River was, according to the Agency, four times more polluted than the second worst Guangdong waterway!

Policies now being implemented may slowly improve the state of Deep Bay; and the Hong Kong–Guangdong Environmental Protection Group is attempting to stem on-going pollution. However, were the pollution to increase, the Nature Reserve's wetlands would become steadily less robust—and unable to attract and feed the present vast numbers of birds.

Deep Bay must contend with still further 'progress', for on both sides of the Shenzhen River there are ambitious plans that take little account of ecological conservation. Development proposals cover no less than 40 per cent of Hong Kong's remaining fish ponds that surround—and at present partly insulate—the Reserve. In Shenzhen there is a relentless momentum to 'develop'.

That the Mai Po Marshes Nature Reserve is one of the most significant wildlife sanctuaries along the South China coast was highlighted by a survey published in February 1995. The study, by Hong Kong University ecologist Steve McChesney, found only about 2,500 birds along a stretch of the eastern Pearl River estuary—compared to some 53,000 birds wintering in Deep Bay.

Yet the bay is hemmed in by development and threatened by pollution. China's Futien Nature Reserve, established in 1984 just across the Shenzhen River, faces the same—or worse—problems.

* *Shenzhen's awesome development is indicated by the growth of its population. In 1980 the permanent population was 84,000; ten years later it was 395,000—with a further 'temporary' population of 614,000.*

Deep Bay's mangrove and mud-flat habitats are of world importance.
But despite this seemingly tranquil setting, its birds and marine life face grim pollution.

There is a clear need to control the future of the Deep Bay area, and to assess carefully the possible impact of every 'development' on its wetlands. Hence, there is an urgent need to enshrine the Nature Reserve with the protection of the internationally recognized Ramsar Convention—to which Hong Kong became an independent signatory in May 1979. At the 1993 Ramsar meeting the Hong Kong government at last stated that the Mai Po marshes were 'worthy' of being named as a Ramsar site—but not until early 1995 did the government approve the actual listing of Mai Po under the Ramsar rules.*

Meanwhile, a golf course and housing development which might severely jeopardize the 'buffer' wetlands surrounding the Nature Reserve was approved by the government's own Town Planning Appeal Board; and four other developers revealed plans that might impact on parts of the Mai Po marshes. Lew Young, the Mai Po Reserve manager, could only restate what had been clear to conservationists for at least a decade:

> *It is imperative that the government urgently develops a comprehensive land-use strategy for the north-western New Territories, in order to prevent piecemeal development and to allow coherent plans for the conservation of fish ponds and other wetland habitats.*

In another article, Lew Young argued that preserving Mai Po 'will require civil servants, developers and conservationists to compromise, put aside traditional differences and sit together to plan for the future of Hong Kong, rather than for their own interests'. He was surely right.

Mai Po symbolizes many Hong Kong conservationists' brightest hopes—and deepest fears. That amid the Territory's dense urban environments a place of such ecological value does still exist often seems remarkable. That many individuals and businesses are willing to support the Nature Reserve's preservation is clear. Yet thus far Mai Po has survived less with the government's enthusiastic support than with its slow, poorly funded, and weakly coordinated recognition. And, for every 'green' ally, Mai Po faces the ecological indifference (or at least ignorance) of Hong Kong's powerful, money-driven developers.

Mai Po, surely, is a litmus test for the future of the Hong Kong countryside. If Hong Kong cannot vigorously act to protect these locally unique, internationally significant wetlands, can there be any hope for other, less remarkable parts of the local landscape?

What, one wonders, would the naturalist G. A. C. Herklots have to say about the threats facing Mai Po today, half a century after he strode across these wetlands? Inside the Mai Po Wildlife Education Centre, close to the laboratories where ecologists now work, is a plaque commemorating Herklots' life: 'A Man for All Seasons' it says. A tireless advocate of nature, Herklots would no doubt urge that we unite to protect Mai Po—and that we seek the inspiration to do so in the country itself.

In his book *The Hong Kong Countryside*, Herklots wrote of the challenges of learning about natural things: 'Having arrived at the heronry, a spot relatively free from excrement is chosen; and, sheltered from the sun by a branch of a tree, we lie down and gaze upwards at the birds.' And he wrote constantly of the pleasures of exploring the natural world:

> *From the bund there is an excellent view of the mangrove swamp and of the mud-flats beyond.... At the [Deep Bay] end of the bund there are a few crab ponds and a dragon boat is embedded in the mud, being dug out each year for the annual races; the prow of this boat is a favourite perch of the Black-capped Kingfisher.*

* *The Ramsar Convention on Wetlands of International Importance Especially as Waterfowl Habitat was established in 1971. Article 3.1 requires that 'Contracting Parties shall formulate and implement their planning so as to promote, as far as possible, the wise use of wetlands in their territory'.*

CONCLUSION

THE FUTURE CHALLENGE

In Kowloon the people mingle with foreigners, and the custom of the residents is that they value money and material things and belittle poetry and studies.

XIN'AN MAGISTRATE, 1847

To live in plenty is not the same as to live well. Living well requires that one feels in harmony with nature, that one knows what to look for: how to find happiness in simple things, in the view of a mountain or the sea.

WILLIAM FULBRIGHT

Today skyscrapers tower below Victoria Peak, where in the 1840s there were only squat stone buildings. But the granite slabs on Victoria Peak are the same that watched over the Hong Kong of Chinese villagers and over the Colony of Hong Kong—and these same rocks, in the years to come, will watch over Hong Kong, the Special Administrative Region. Protruding from rugged slopes, these granites have barely changed in centuries.

Hong Kong's urban canyons are testimony to human enterprise and energy. Its rural peaks and valleys are testimony to the ingenuity and power of nature. They stretch back to the dawn of time and reflect the origins of life. Year after year, if left alone, the countryside's plants and creatures will repeat ancient and complex cycles.

However, large parts of the Territory's remaining countryside will be lost unless protected with the utmost vigilance, unless nature is seen to have a value that transcends dollars, unless enough people insist that the Country Parks are left untouched forever.

It was the summer of 1994. Guangdong and Hong Kong had endured weeks of rain, but now the sky was cerulean. Beyond Kowloon cumulus clouds towered over majestic peaks. The harbour, blue-green, criss-crossed with white wakes, was shimmering.

Near Central, however, slicks of caramel silt spread across the water as—dredger-load by dredger-load—an extensive reclamation took shape. Across the harbour, beside Kowloon, another vast reclamation had almost been completed.

How would these controversial reclamations finally appear, I wondered? Would they be crowded with clashing, densely packed high-rises—or would the official vision of civilized, spacious places be realized? Would still further reclamations transform the harbour into a virtual channel—or would wisdom prevail and control their extent? The ultimate appearance of these harbour reclamations might also indicate the fate of Hong Kong's wild places. For if those who govern the Territory disregard its urban environment, they are unlikely to concern themselves with the preservation of its countryside.

Such were my thoughts as I passed through Central one day. I had begun the day on a Lamma Island hilltop, photographing from there Hong Kong Island's southern side as dawn revealed its grandeur. Now, revelling in a few days of unusually clear light, I was en route to Bride's Pool, near Plover Cove Reservoir.

I reached Bride's Pool about midday. My last visit had been in winter: then the hills were brown, the streams shallow. Now the country was lush: the hills were emerald, the streams in spate. Where a dry-season fire had blackened the ground, waist-high grasses now grew.

Under an intense blue sky I followed a valley down towards Double Haven, past hillsides seeping water from recent rain. The country, enveloped in smells of summer growth, was alive with butterflies. The further I walked the more remote—and superficial—Hong Kong's urban life seemed. As I crouched to drink from a spring, luxuriating in its crystal water, nature washed the city's values away.

Beyond Bride's Pool the region is virtually uninhabited. One can walk for days and see barely a soul. This is country not to be trifled with: the paths are often overgrown, hillsides plunge suddenly into dank forested gullies.

There is a palpable sense of the past, in the land itself and in its abandoned homes. Pine roof-beams lie collapsed and rotting, ceramic urns overflow with rain water, rattan baskets disintegrate under spider web canopies, ferns encroach on moss-green walls.

Hong Kong is much more than the mere creation of the last 150-odd years, and its future stretches much further than the next generation or two. Wisdom thus demands that we accept the limits of nature, move away from our obsession with 'growth', devise a sustainable city, and preserve the surrounding countryside.

In the past, peasants exploited the country mostly out of necessity. Though they had scant knowledge of long-term ecological change, they still enriched the land to some extent. Today, much of our exploitation of nature reflects a desire for wealth. Despite abundant scientific evidence concerning the dire consequences of environmental degradation, we continue to abuse the land.

In the 1950s Hong Kong committed itself to a sustained programme of rehousing, because, as a landmark government report stated: 'the spectacle of such extremes of misery, need, and danger in the heart of a prosperous community could no longer be tolerated'.

The Territory's subsequent remarkable achievements, its economic and social revolution of the 1960s and 1970s, hint at the potential for an environmental revolution today. But this will only occur when a widespread resolve to 'no longer tolerate' Hong Kong's environmental woes is born—when Hong Kong people unite to transcend old attitudes and habits—unite to forge a new 'green' consciousness, a new environmental ethic.

Many other places have little immediate hope of restoring their ravaged surroundings. But affluent Hong Kong, were it to harness the competitive energy, resolve, and skill that transformed its post-war society, has the resources, technology, and wealth to launch a sustained environmental agenda: to 'green' its urban areas, to preserve its countryside, and to help reverse the degradation of the entire Pearl River delta.

However Hong Kong also has the power to lay waste its surroundings—literally to recast the natural landscape until the remaining countryside becomes a mere adjunct to greedy 'development'.

Choices must be made. Whether the countryside will be preserved or degraded depends on considered values—and on government, business, community, and individual decisions. Is Hong Kong to become a place where edifices eclipse their natural setting? Or will it become a model international city whose wealth enhances nature?

Those who say that Hong Kong's natural beauty has been destroyed already are wrong, but so too are the apologists who choose to ignore its environmental decline. As Edmund Burke wrote: 'Nobody made a greater mistake than he who did nothing because he could do only a little'.

Some twelve centuries ago, in a time of dynastic upheaval in China, Du Fu wrote:

The state is shattered;
Mountains and rivers remain.

In the years to come, if Hong Kong people allow—or worse still encourage—rash 'development' to continue or to spread in from Guangdong, the Territory's mountains might be shattered—by technological power and rapacious, short-term values. Were that to happen, future Hong Kong people might decry the folly. For, in degrading nature, we also degrade ourselves—and so, finally, 'the state'.

The year 1997 will pass unnoticed by the timeless natural landscape—by its mountains, streams, and coasts, by its flora and fauna. But what will the historic events of that year bring to the countryside in the longer term?

Will Hong Kong people continue to exploit their natural landscape for short-term gain—or, grasping the challenge of responsibility that 1997 brings, will they resolve to protect their surroundings? In the decades to come, will posterity judge those who guide Hong Kong's post-1997 destiny as 'custodians' or 'developers'—leaders mindful of the generations to come, or leaders concerned only with today?

Hong Kong's man-made skyline, icon of its economic success, impresses and dazzles, brilliant as much as beautiful. Hong Kong's natural landscape also impresses—and, majestic and beautiful, it delights and soothes. Both deserve our respect: the future well-being of both lies in our hands.

Further Reading

This book draws on widely different disciplines: ecology, geography and the 'earth' sciences; history, literature and the visual arts; serious and popular writing. It includes no 'original' research, but I hope it presents what is known in a fresh context. Some of the sources are well known, others are little known—and some are almost forgotten.

Accessible sources dealing with Hong Kong's pre-colonial ecology are extremely few. What exists is very general and does not quantify environmental change. Post-1841 British records, especially those concerning flora and reforestation, are much more numerous and detailed. As these and other nature or travel writings are personally and politically 'neutral', there seems no reason to doubt their authenticity.

Four key books provided an invaluable framework for my more discursive rambles around the country. These books together provide an excellent introduction to Hong Kong's ecology and geography. They are (in alphabetical order): the geography of Hong Kong edited by T. N. Chiu and C. L. So; David Dudgeon and Richard Corlett's ecological study of Hong Kong; G. A. C. Herklots' exploration of the local countryside; and Stella Thrower's overview of the Country Parks.

The Hong Kong government's annual reports include interesting details concerning human and natural impacts on the local environment. Those issued before 1941, which appeared as pamphlets, are useful though brief. The post-war reports vividly chronicle Hong Kong's later transformation.

The annual reports of the Environmental Protection Department helped with the last chapters, as did various recent reports from the Planning, Environment and Lands Branch. Almost all the statistics in those chapters are from these reports. The journals of Hong Kong's various environmental groups were also helpful.

The wide range of booklets published by the Urban Council on aspects of Hong Kong's ecology gave useful background. *The Hong Kong Naturalist*, published during the 1930s, has much of interest. The journals of the Hong Kong Branch of the Royal Asiatic Society provide diverse and revealing insights, and some recent academic papers helped.

References

Where material comes from a single source, the author or title is mentioned in the text. Government statements and statistics, unless otherwise noted, are from the annual reports.

BIBLIOGRAPHY

Allom, T. & Wright, G. N., *China in a Series of Views*, London: Fisher, 1843.

Atherton, M. J. & Burnett, A. D., *Hong Kong Rocks*, Hong Kong: Urban Council, 1986.

Bard, S., *In Search of the Past*, Hong Kong: Urban Council, 1988.

Bentham, G., *Flora Hongkongensis: A Description of the Flowering Plants and Ferns*, London: Lovell Reeve, 1861.

Bird, I. L., *The Golden Chersonese*, Kuala Lumpur: Oxford University Press, 1967.

Blunden, E., *A Hong Kong House: Poems 1951–1961*, London: Collins, 1962.

Braudel, F., *The Mediterranean and the Mediterranean World in the Age of Phillip II*, London: Collins, 1972.

Bristow, R., *Land-use Planning in Hong Kong*, Hong Kong: Oxford University Press, 1987.

Chan, C., *Hong Kong After Forty Years' Vicissitudes*, Hong Kong: Joint Publishing Co, 1985.

Chan, C. & Hills, P., *Limited Gains, Grassroots Mobilization and the Environment in Hong Kong*, Hong Kong: University of Hong Kong, 1993.

Chin, P. C., *The Life History of a Tropical Cyclone*, Hong Kong: Royal Observatory, 1977.

Chiu, T. N. & So, C. L. (eds.), *A Geography of Hong Kong*, Hong Kong: Oxford University Press, 1986.

Coates, A., *A Mountain of Light*, Hong Kong: Heinemann Educational Books Ltd., 1977.

Coates, A., *Myself A Mandarin*, Hong Kong: Oxford University Press, 1987.

Crook, A. H., *The Flowering Plants of Hong Kong*, Hong Kong: Ye Olde Printerie Ltd., 1930.

Da Silva, A., *Tai Yu Shan: Traditional Ecological Adaption on a South Chinese Island*, Taipei: Orient Cultural Service, 1972.

Des Voeux, W., *My Colonial Service*, 2 vols., London: John Murray, 1903.

Dudgeon, D. & Corlett, R., *Hills and Streams: An Ecology of Hong Kong*, Hong Kong: Hong Kong University Press, 1994.

Dunn, S. T. & Tutcher, W. J., *Flora of Kwangtung and Hong Kong*, London: His Majesty's Stationery Office, 1912.

Dury, G. H., *The Face of the Earth*, London: Allen & Unwin, 1986.

Dyson, A., *From Time Ball to Atomic Clock: A History of the Royal Observatory*, Hong Kong: Government Information Services, 1983.

Eitel, E. J., *Europe in China: The History of Hong Kong from the Beginning to the Year 1882*, Hong Kong: Kelly & Walsh, 1895.

Empson, H., *Mapping Hong Kong: A Historical Atlas*, Hong Kong: Hong Kong Government, 1992.

Endacott, G. B., *A History of Hong Kong*, Hong Kong: Oxford University Press, 1958.

Faure, D., *The Structure of Chinese Rural Society*, Hong Kong: Oxford University Press, 1986.

Finn, D. J. & Ryan, T. F., *Archaeological Finds on Lamma Island*, Hong Kong: Ricci, 1958.

Fisher, F., *Control of the Environment in Hong Kong*, London: Environmental Resources Ltd, 1975.

Fortune, R., *Three Years Wandering in China*, London: John Murray, 1847.

Gee, R., *Rambles in Hong Kong*, Hong Kong: Oxford University Press, 1992.

Gleason, G., *Hong Kong*, London: Robert Hale Ltd, 1964.

Grantham, A., *Via Ports*, Hong Kong: Hong Kong University Press, 1965.

Griffiths, D. A., *Grasses and Sedges of Hong Kong*, Hong Kong: Urban Council, 1983.

Hayes, J., *The Rural Communities of Hong Kong*, Hong Kong: Oxford University Press, 1983.

Hayes, J., *Tsuen Wan: Growth of a 'New Town'*, Hong Kong: Oxford University Press, 1993.

Herklots, G. A. C., *The Hong Kong Countryside*, Hong Kong: South China Morning Post Ltd, 1951.

Heywood, G. S. P., *Rambles in Hong Kong*, Hong Kong: Kelly & Walsh, 1951.

Hill, D. & Phillipps, K., *Hong Kong Animals*, Hong Kong: Hong Kong Government, 1981.

Hill, D. S., Gott, B., Morton, B. S. & Hodgkiss, I. J., *Hong Kong Ecological Habitats, Flora and Fauna*, Hong Kong: University of Hong Kong, 1978.

Hodgkiss, I. J., Thrower, S. L. & Man, S. H., *Introduction to the Ecology of Hong Kong*, Vols 1 and 2, Hong Kong: Federal Press, 1981.

Hong Kong Land Investment Co., *In Far Eastern Waters*, Hong Kong: Ye Olde Printerie Ltd., 1930.

Hsia, J. & Lai, T. C., *Hong Kong: Images on Shifting Waters*, Hong Kong: Kelly & Walsh, 1977.

Hurley, R. C., *Picturesque Hong Kong and Dependencies*, Hong Kong: Commercial Press Ltd., 1925.

Irving, R. & Morton, B., *A Geography of the Mai Po Marshes*, Hong Kong: Hong Kong University Press, 1988.

Jarrett, V. H. C., *Familiar Wildflowers of Hong Kong*, Hong Kong: South China Morning Post, 1937.

Jim, C. Y., *Trees in Hong Kong: Species for Landscape Planting*, Hong Kong: Hong Kong University Press, 1990.

Jim, C. Y., *Wan Chai Green Trail*, Hong Kong: Conservancy Association, 1994.

Kemp, D., *Twelve Hong Kong Walks*, Hong Kong: Oxford University Press, 1985.

Kershaw, J. C., *Butterflies of Hong Kong*, Hong Kong: Kelly & Walsh, 1907.

Knight, M. C., *Hong Kong Journal 1949*, Oxford: Classic, 1950.

Lo, H. L., *Hong Kong and Western Cultures*, Tokyo: East Asian Cultural Studies, 1963.

Lo, H. L., *The Role of Hong Kong in the Cultural Interchange Between East and West*, Tokyo: East Asian Cultural Studies, 1963.

Luff, J., *Hong Kong Cavalcade*, Hong Kong: South China Morning Post Ltd, 1968.

Meacham, W., *Archaeology in Hong Kong*, Hong Kong: Heinemann, 1980.

Morres, A. (ed.), *Another Hong Kong*, Hong Kong: Emphasis Ltd, 1989.

Morton, B. (ed.), *The Future of the Hong Kong Seashore*, Hong Kong: Oxford University Press, 1979.

Morton, B. & Ruxton, J., *Hoi Ha Wan*, Hong Kong: World Wide Fund for Nature, 1992.

Ng, P. Y. L., *New Peace County: A Chinese Gazetteer of the Hong Kong Region*, Hong Kong: Hong Kong University Press, 1983.

Nutt, T., Bale, C. & Ho T., *The MacLehose Trail*, Hong Kong: The Chinese University Press, 1992.

Ommanney, F. D., *Fragrant Harbour: A Private View of Hong Kong*, London: Hutchison & Co Ltd, 1962.

Orange, J., *The Chater Collection: Pictures of Hong Kong and Macau*, London: Butterworth, 1924.

Phillips, R. J., *Kowloon–Canton Railway (British Section): A History*, Hong Kong: The Urban Council, 1990.

Ponting, C., *A Green History of the World*, London: Penguin Books, 1991.

Pope Hennessy, J, *Half Crown Colony: A Hong Kong Notebook*, London: Jonathan Cape, 1969.

Rand, C., *Hong Kong: The Island Between*, New York: Alfred Knopf, 1952.

Rayne, R. (ed.), *The White Pony: An Anthology of Chinese Poetry*, New York: New American Library, 1960.

Rodwell, S., *Historic Hong Kong*, Hong Kong: The Guidebook Company, 1991.

Ryan, T. F., *The Story of a Hundred Years: The Pontifical Institute of Foreign Missions in Hong Kong 1858–1958*, Hong Kong: Catholic Truth Society, 1959.

Sayer, G. R., *Hong Kong 1841–1862*, Hong Kong: Hong Kong University Press, 1937.

Sayer, G. R., *Hong Kong 1862–1919*, Hong Kong: Hong Kong University Press, 1975.

Schepel, K., *Magic Walks*, Hong Kong: Alternative Press, 1990

Sinclair, K. (ed.), *A Soldier's View of Empire: The Reminiscences of James Bodell 1831–1892*, London: The Bodley Head, 1982.

Stericker, J. & V. S., *Hong Kong in Picture and Story*, Hong Kong: Tai Wah Press, 1953.

Stokes, G. G. & Stokes, J., *Queen's College, Its History 1862–1987*, Hong Kong: Queen's College Old Boys' Association, 1987.

Tang, H. C., *Hong Kong Trees*, Hong Kong: Urban Council, 1969

Thompson, J., *China, the Land and its People*, London: Warner, 1977.

Thompson, J., *Illustrations of China and its People*, London: Sampson Low, 1873.

Thoreau, H. D., *The Maine Woods*, Princeton: Princeton University Press, 1972.

Thrower, S. L., *Hong Kong Country Parks*, Hong Kong: Hong Kong Government, 1984.

Tuan, Y. F., *Space and Place: The Perspective of Experience*, Minnesota: University of Minnesota Press, 1977.

Tuan, Y. F., *The World's Landscapes—China*, London: Longman Limited, 1970.

Turner, J. A., *Kwang Tung or Five Years in South China*, London: S. W. Partridge, 1894.

Van Slyke, L. P., *Yangtze: Nature, History and the River*, Massachusetts: Addison-Wesley Publishing, 1988.

Viney, C., Phillipps K. & Lam, C. Y., *Birds of Hong Kong and South China*, Hong Kong: Government Printer, 1994.

Walden, B. M. & Hu, S. Y., *Wild Flowers of Hong Kong Around the Year*, Hong Kong: Sino-American Publishing Company, 1977.

Williams, M. & Pitts, M., *The Green Dragon: Hong Kong's Living Environment*, Hong Kong: Green Dragon Publishing, 1994.

Wise, M. (ed.), *Travellers' Tales of the South China Coast*, Singapore: Times Books, 1986.

Yeung, Y. M. & Chu, K. Y., *Guangdong: Survey of a Province Undergoing Rapid Change*, Hong Kong: Chinese University Press, 1994.

Young, L. & Melville, D. S., 'Conservation of the Deep Bay Environment' in *The Marine Biology of the South China Sea*, Hong Kong: Hong Kong University Press, 1993.

Hiking and Conservation Notes

I hope these notes will encourage people unfamiliar with hiking or Hong Kong to venture into the Country Parks' more remote areas. The local countryside deserves respect, both for one's own safety and for its conservation.

Maps

Hiking in Hong Kong without good maps is not advisable. The government's *Countryside Series* maps are ideal. Their 1:25,000 scale gives a detailed picture of each region, and they have safety notes, clear contours, and graded tracks. The *Countryside Series* maps also provide information that helps one appreciate the country. They are available from the Government Bookshop at 66 Queensway.

Make sure you can orient a map and read the contour lines to plan a route. With Hong Kong's steep terrain, contour heights, not horizontal distances, are the best measure of the time and energy a route will demand.

Another excellent map is the World Wide Fund for Nature's *Hong Kong Vegetation Map* (1993). This two-sheet 1:50,000 map of the Territory, and its explanatory booklet, give a very detailed picture of vegetation types and areas. Besides its intrinsic interest, the map is invaluable to hikers, who need to know whether country is woodland, shrubland, or grassland.

Hiking and Safety

The proximity of Hong Kong's countryside to the urban areas can easily encourage a false sense of security. However, approached carelessly, the local countryside is hazardous. A few fatalities occur most years in the Country Parks. Adequate preparation and cautious hiking are the best safeguards. Keeping the following in mind should minimize the risks and increase your enjoyment.

- Hiking in Hong Kong can be exhausting, especially in summer. Plan your route beforehand, allowing for the terrain, weather, and your party's weakest member—and check the weather forecast before you set off.
- Always have a good map and a reliable compass. Take ample food and water, and carry a basic medical kit, water purifying tablets, a knife, and a torch.
- Wear a whistle around your neck. If injured, you can attract attention with a whistle long after you can no longer shout. A small mirror can be used for 'flashing' your location to rescuers or aircraft. A 'space blanket' is invaluable, both for warmth and to make your position visible from the air.
- Even when hiking in a group *always leave details* of your general route and expected return time with someone who can raise the alarm if need be. Searching for people without an area to start from is almost impossible.
- Be cautious if you come to an overgrown path. It may be better to find another route, even if it is longer. But do not be put off too easily; small areas of shrubs often invade paths, and, if you can get through, the path may resume further on. Keep an eye on the middle distance to see where a path is leading, or if it disappears. Beware of old hollows, including empty graves, in overgrown areas.
- If the weather becomes excessively hot, cold, rainy, or misty, adapt your plans to suit. If visibility fails it is far safer to stay put than to push on: unseen hazards can be fatal, a bleak night is soon forgotten. A small radio is useful for forecasts (and is calming).
- If you are forced to camp out without overnight gear, keep as warm—or cool—as the weather dictates. Build a windbreak or shelter, nestle into low shrubs, find a hollow and cover yourself with leaves. Large rubbish-bin liners make ideal emergency sleeping bags to keep off wind chill.
- Unless some danger demands instantaneous action, if in trouble *always stop*. Rest, relax, slowly assess your situation, make a clear decision, then act on it.

Hiking Alone

Lone hikers are particularly vulnerable in Hong Kong's rugged country—and, if injured, extremely difficult to locate. If hiking alone the precautions above should be strictly adhered to.

Above all, if visibility fails, *stay put*. In thick mist I was once tempted to take a sidetrack down off Pat Sin Leng, marked on the map but almost invisible in the mist. However, caution fortunately said 'don't'. I was told later that the 'short cut' was in fact overgrown; someone had recently fallen off it and been killed.

When hiking alone it is essential to have a contact who knows one's expected route and return time. Even on 'easy' hikes it is always best to have a contact, as one can be led off 'safe' tracks onto hazardous ones.

While walking around Tai Tam valley, without having advised my contact, I decided to climb to Ma Kong Shan (The Twins, 386 metres). I had never been there before. The track winding tortuously through dense trees, was narrow, rocky, and extremely steep. Coming downhill, balancing my camera gear and tripod, I had to swing from branch to branch. Had I fallen badly (as I almost did), and been immobilized, I would have been in an unpleasant and risky predicament—the track was clearly rarely used.

CONSERVING THE COUNTRYSIDE

Fires are the greatest threat to Hong Kong's countryside. Lighting them is illegal except in designated areas. Certainly never light fires on windy days or near dry vegetation; never leave fires unattended; and always ensure that fires are properly extinguished.

If you see a small fire try to extinguish it. Beat out the flames with leafy branches, clothes, or anything that can smother them. If you see a large fire, telephone the emergency services.

Pitch tents on barren ground to avoid crushing plants. If you gather grasses for a soft bed try to use dead material; and avoid cutting vegetation to 'improve' a campsite. Wash away from streams and bury excreta in shallow holes. In remote areas carry your rubbish out, to avoid filling bins which are emptied only occasionally. If you can, collect some litter on your way home.

PHOTOGRAPHIC NOTES

The major challenge in landscape photography is knowing where to photograph—and predicting when to be there. A compass for predicting light directions and an alarm clock for early starts are stock-in-trade. Strong vantage points, together with optimum weather and light, are the 'secret' behind most of the photographs. Pressing the shutter is almost an afterthought.

Hong Kong's changeable weather meant that many of my excursions were abortive. Air pollution is another difficulty, as it often greatly reduces light penetration and clarity. In late afternoons and early mornings—when elsewhere colours, details, and textures are usually at their strongest—Hong Kong's light often fades away behind a wall of pollution.

To cut down on weight I mostly went on two-day overnight excursions. I camped with the bare essentials, and no tent, but still usually carried about twenty kilograms—mostly camera gear. As well as the usual photographic and safety equipment, I took cotton gloves and lengths of cord and rope—for scrambling up rocks, hauling up and lowering gear, tying down branches to give clear views, and, occasionally, for anchoring myself to trees or rocks when photographing in risky positions.

I use Olympus OM4-Ti or OM4 bodies, with Olympus Zuiko lenses, and Fuji Velvia 50 ASA transparency film. The Olympus cameras' 'spot metering' is particularly valuable given Hong Kong's often rapidly changing light, but I always bracket exposures. The lenses I most often use are: 24 mm, 50 mm macro, 100 mm, 200 mm, 35–70 mm, and 75–150 mm. I almost always use a tripod, except with the 24 mm, to maximize depth-of-field and definition by 'stopping-down'—usually at least to f8, and often to f11 or f22. I use polarizing filters to reduce glare and increase colour saturation, but no 'colour-adjusting' filters.

My camera equipment is sponsored by R. Gunz Ltd, the Olympus agent in Australia. My sincere thanks go to Gunz, and especially Bob Pattie, for supporting my work over a number of years.

In Hong Kong my photographic supplies come from Photo Scientific Appliances Ltd, in Stanley Street, Central—an oasis of friendly and professional assistance. My special thanks go to Mr Poon Ka-Kui, Mr Tony Cheng, and staff.

INDEX

GUANGDONG
SHENZHEN
Sha Tau Kok
Yim Tso Ha
Lo Wu
DEEP BAY
PEARL RIVER ESTUARY
SHEKOU
SHEUNG SHUI
FANLING
Sha Lo Tung
Pat Sin Leng
TIN SHUI WAI
Ping Shan
YUEN LONG
Kam Tin
TAI PO
Shek Kong
Ng Tung Chai Waterfall
Grassy Hill 647m
Tai Mo Shan 957m
Shing Mon Reservoir
SHA TIN
Tai Lam Chung Reservoir
Castle Park 583m
TUEN MUN
TSUEN WAN
Tate's 577m
TSING YI
KWAI CHUNG
Beacon Hill 452m
SHA CHAU
KAI TAK
Chek Lap Kok Airport Site
Yau Ma Tei
KOWLOON
VICTORIA HARBOUR
Discovery Bay
Tung Chung
LANTAU ISLAND
PENG CHAU
Green Island
Kennedy Town
Victoria Peak 552m
Mui Wo
SUNSHINE ISLAND
Wan Chai Gap
Violet Hill 433m
Ngong Ping
Lantau Peak 934m
Sunset Peak 869m
Tai O
HEI LING CHAU
Aberdeen
EAST LAMMA CHANNEL
Shek Pik Reservoir
Shek Pik
Chi Ma Wan Peninsula
Yung Shue Wan
Repulse Bay
Chung Hom Kok
CHEUNG CHAU
LAMMA ISLAND
Fan Lau Kok
LANTAU CHANNEL
SHEK KWU CHAU
Mount Stenhouse 353m
Sham Wan
SOKO ISLANDS
km 0 2 4 6 8 10 12 14 km